P9-DZY-354

GREAT DISASTERS

FIRES

Ana María Rodríguez, *Book Editor*

Bonnie Szumski, *Publisher*
Scott Barbour, *Managing Editor*

**GREENHAVEN
PRESS®**

THOMSON

™

GALE

San Diego • Detroit • New York • San Francisco • Cleveland
New Haven, Conn. • Waterville, Maine • London • Munich

THOMSON

GALE

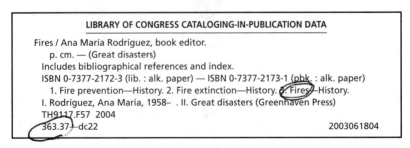

LIBRARY OF CONGRESS CATALOGING-IN-PUBLICATION DATA

Fires / Ana María Rodríguez, book editor.
   p. cm. — (Great disasters)
  Includes bibliographical references and index.
  ISBN 0-7377-2172-3 (lib. : alk. paper) — ISBN 0-7377-2173-1 (pbk. : alk. paper)
   1. Fire prevention—History. 2. Fire extinction—History. 8. Fires—History.
I. Rodríguez, Ana María, 1958– . II. Great disasters (Greenhaven Press)
TH9117.F57 2004
363.37—dc22
                                             2003061804

Printed in the United States of America

CONTENTS

what was then America's fourth largest city. The fire
was such a devastating chapter in the city's history
that Chicagoans refer to events as happening "before"
or "after" the Chicago fire.

Chapter 3: Fire Prevention and Damage Control

left forests choked with fuel, it has resulted in even more disastrous conflagrations.

H umans have an ambivalent relationship with their home planet, nurtured on the one hand by Earth's bounty but devastated on the other hand by its catastrophic natural disasters. While these events are the results of the natural processes of Earth, their consequences for humans frequently include the disastrous destruction of lives and property. For example, when the volcanic island of Krakatau exploded in 1883, the eruption generated vast seismic sea waves called tsunamis that killed about thirty-six thousand people in Indonesia. In a single twenty-four-hour period in the United States in 1974, at least 148 tornadoes carved paths of death and destruction across thirteen states. In 1976, an earthquake completely destroyed the industrial city of Tangshan, China, killing more than 250,000 residents.

Some natural disasters have gone beyond relatively localized destruction to completely alter the course of human history. Archaeological evidence suggests that one of the greatest natural disasters in world history happened in A.D. 535, when an Indonesian "supervolcano" exploded near the same site where Krakatau arose later. The dust and debris from this gigantic eruption blocked the light and heat of the sun for eighteen months, radically altering weather patterns around the world and causing crop failure in Asia and the Middle East. Rodent populations increased with the weather changes, causing an epidemic of bubonic plague that decimated entire populations in Africa and Europe. The most powerful volcanic eruption in recorded human history also happened in Indonesia. When the volcano Tambora erupted in 1815, it ejected an estimated 1.7 million tons of debris in an explosion that was heard more than a thousand miles away and that continued to rumble for three months. Atmospheric dust from the eruption blocked much of the sun's heat, producing what was called "the year without summer" and creating worldwide climatic havoc, starvation, and disease.

As these examples illustrate, natural disasters can have as much impact on human societies as the bloodiest wars and most chaotic political revolutions. Therefore, they are as worthy of study as the major events of world history. As with the study of social and political events, the exploration of natural disasters can illuminate the causes of these catastrophes and target the lessons learned about how to mitigate and prevent the loss of life when disaster strikes again. By examining these events and the forces behind them, the Greenhaven Press Great Disasters series is designed to help students better understand such cataclysmic events. Each anthology in the series focuses on a specific type of natural disaster or a particular disastrous event in history. An introductory essay provides a general overview of the subject of the anthology, placing natural disasters in historical and scientific context. The essays that follow, written by specialists in the field, researchers, journalists, witnesses, and scientists, explore the science and nature of natural disasters, describing particular disasters in detail and discussing related issues, such as predicting, averting, or managing disasters. To aid the reader in choosing appropriate material, each essay is preceded by a concise summary of its content and biographical information about its author.

In addition, each volume contains extensive material to help the student researcher. An annotated table of contents and a comprehensive index help readers quickly locate particular subjects of interest. To guide students in further research, each volume features an extensive bibliography including books, periodicals, and related Internet websites. Finally, appendixes provide glossaries of terms, tables of measurements, chronological charts of major disasters, and related materials. With its many useful features, the Greenhaven Press Great Disasters series offers students a fascinating and awe-inspiring look at the deadly power of Earth's natural forces and their catastrophic impact on humans.

On October 8, 1871, the deadliest fire in American history hit the lumber town of Peshtigo, Wisconsin. The fire burned more than two thousand square miles of woods, fields, and settlements and killed nearly two thousand people. The conflagration began as scattered fires slowly burned sections of the forests around Peshtigo and produced a thick layer of smoke over the town. For days people covered their mouths with handkerchiefs to protect themselves from the smoke and became increasingly aware of the smell of burning sap. Despite these warning signs of a disaster in the making, however, most people went about their daily routines.

As strong winds from a massive cold front reached Peshtigo, the fires exploded. Fanned by strong drafts, the blazes developed into a tornado of fire more than one thousand feet high and five miles wide. The conflagration rushed through town and forests, incinerating people, property, and natural resources. Author and firefighter Peter M. Leschak describes the disaster in his book *Ghosts of the Fireground: Echoes from the Great Peshtigo Fire and the Calling of a Wildland Firefighter:*

> Nails fused into an iron glob by unimaginable heat; an oxygen-sucking firestorm that was probably unequaled in power until the incendiary bombing of Dresden or the destruction of Hiroshima; a panicked woman plunging into the frigid Peshtigo River with a bundled infant only to discover the wrappings were empty; a man running to relative safety with his wife on his back, realizing too late that he had the wrong woman.

Despite being a catastrophe of unprecedented magnitude, the Peshtigo fire has remained virtually unknown to most Americans, primarily because it occurred on the same day as the Great Fire of Chicago (Chicago is located about 260 miles south of Peshtigo). Few knew that the remote, small town of Peshtigo, with a population of two thousand people, had practically vanished from existence until Peshtigo survivors left the disaster area searching for help. In contrast, news that a gigantic fire had killed about three hundred people and consumed Chicago, a prosperous city with one hundred thousand residents, quickly spread throughout the nation. As Leschak explains, "Chicago grabbed the first headlines, delaying aid to the bedeviled victims of Peshtigo."

Not surprisingly, large urban fires, because they occur in well-known cities and can cause millions of dollars in damage and high human casualties, garner substantial public notice. Wildfires, however, are equally destructive from a scientific point of view.

Both urban and rural fires derive their enormous power of destruction from a precise combination of fuel, oxygen, and heat. In a matter of days, flames can destroy hundred-year-old forests, magnificent cities, and countless lives. Fires can also heat the surrounding air to temperatures so high that people's lungs burn even before the fire itself reaches them. In addition, most fires release large amounts of smoke, which can kill people. Smoke inhalation causes death by asphyxiation, one of the most common causes of injury and death during fires.

Urban and wildland fires start when a source of heat increases a fuel's temperature high enough for the fuel to burst into flames. Lightning is the most frequent natural source of heat that ignites wildfires, but careless people are increasingly causing fires. Fire experts claim that losses of property and natural resources as a result of human-caused wildfires are lately surpassing the losses due to lightning-caused fires. Historically, most fires in urban areas have been started by human carelessness, such as when people leave heating appliances unattended or drop smoldering cigarette butts into couch cushions.

Some fires will always occur, but fire experts emphasize the importance of fire prevention to minimize the number of fires that occur every year. Many urban fires would be avoidable if people became aware of the fire hazards posed by heating devices, for example. The public fire prevention campaign started by Smokey Bear in 1944 with the slogan "Only YOU Can Prevent Forest Fires" succeeded in raising people's awareness of how wildland fires are started. Wildland fire managers believe this campaign helped reduce the number of acres burned from more than 22 million per year in the 1940s to 9 million acres per year in the 1950s. Many forest managers also contend that cutting down dying trees and clearing forest undergrowth can prevent wildfires, although environmentalists argue that such claims simply provide excuses to loggers to destroy the nation's forests.

To mitigate the damage caused by urban and wildland fires, much effort has been devoted to early fire detection. Fires are easier to control when they are detected early, before they have released large amounts of heat and smoke. Firefighters can approach small fires without endangering themselves and can put them out quickly before they grow into uncontrollable conflagrations. Fire detection in the wildland occurs on the ground (by rangers walking rounds in forests or observing from lookout towers), from the air (airplane personnel looking for smoke or using infrared cameras to detect heat released from fires), and from space (using satellites to track the progress of fires). In the city the use of smoke detectors in homes and other buildings has proven effective in detecting fires early, thereby saving lives and property.

Despite all the prevention and early detection efforts, fires are common all over the world. Fighting fires has become the job of an elite group of men and women specially trained to keep small fires from developing into uncontrollable blazes. Hotshots, who fight wildland fires, and urban firefighters have one of the most dangerous jobs in the world; they risk their lives every day to save people and property. Scientists have found that firefighters on the job expend as much energy as soldiers in combat. Hotshots in action, for example, spend about seven hours daily in constant physical activity carrying seventy pounds of equipment, including a chain saw, fuel, and tools. They wear bulky fire-resistant pants and shirt and endure heat, cold, smoke, and the thin air of high elevations. Sometimes firefighters have to literally run for their lives if the wind changes and pushes the fire toward them.

Even though fires have an enormous destructive power, they can also bring benefits. Fires clean the forest floor by consuming dead vegetation, recycling the nutrients back into the soil. Fires also burn the forest canopy, allowing sunlight to shine on seedlings growing on the forest floor. In fact, shortly after most wildfires, plants sprout more quickly and grow stronger thanks to soil enriched with ashes left by the fire. When a fire burns a city, it usually provides an opportunity for people to rebuild, incorporating into the new structures modern warning devices and fire-resistant building materials that reduce future fire danger. Indeed, the challenge of present and future generations is not to achieve the impossible task of avoiding fires completely, but to learn to live with them by minimizing human and material losses.

The Science and Study of Fires

The Global Fire Scene

BY THE ASSOCIATION OF SOUTHEAST ASIAN NATIONS RESPONSE STRATEGY TEAM

Fire experts are alarmed by the increasing incidence and intensity of wildfires worldwide and the smoke and haze they cause. In the following selection, excerpted from their report Fire, Smoke, and Haze, *the Association of Southeast Asian Nations (ASEAN) Response Strategy Team discusses recent wildfires, including the 1983 Ash Wednesday Fire in Australia that killed seventy-six people, three hundred thousand sheep, and burned more than twenty-five hundred homes. The team also describes the 1982–1983 fire in Côte d'Ivoire in West Africa that killed more than one hundred people and consumed tens of thousands of acres of cocoa and coffee plantations, which drastically impacted the local economy.*

The fire experts also discuss both the human and natural causes that have triggered these massive wildfires, which consume millions of acres of the world's forests every year. The experts conclude that one reason wildfires are catastrophic today is because people are increasingly using fire to clear land for agricultural use and often lose control of the fire. Moreover, recent severe droughts and persistent high temperatures have severely dried soil and vegetation, creating conditions conducive to wildfires. The team argues that wildfires are a global concern because the haze and smoke from the fires are slowly spreading throughout the world and adversely affecting the environment, the world economy, and human health. In response to these concerns, more effective information-gathering tools are being developed that can help scientists better predict and manage wildfires.

The ASEAN team is formed by fire experts from Southeast Asian nations who share experiences, information, responsibilities, and benefits while working toward the common good of the region. The team responds to fire and haze disasters by coordinating the efforts in the affected areas to mitigate the disasters' adverse effects.

The Association of Southeast Asian Nations Response Strategy Team, "Understanding Forest Fires—The Global Fire Scene," *Fire, Smoke, and Haze: The ASEAN Response Strategy*, edited by S. Tahir Qadri. Manila, Philippines: Asian Development Bank, 2001. Copyright © 2001 by Asian Development Bank. Reproduced by permission.

T he Report on Global Environmental Outlook 2000 prepared by the United Nations Environment Programme (UNEP) paints a devastating picture of the earth's health at the dawn of the new millennium. The planet is undergoing an unsustainable course of development, fueled by a relentless decline in the environment and degradation of natural resources.

Contributing to this decline are the uncontrolled wildfires rampaging through lands and forests, affecting the environmental quality and ecological resilience of our habitat. Occurring in agricultural land, forests, and rural areas, spreading from one area to another, burning furiously, and causing heavy and suffocating haze, the fires that ravaged the Association of Southeast Asian Nations (ASEAN) region in 1997–1998 reached disastrous proportions. The environmental, economic, and social dimensions and impact of the catastrophic fires, and the associated transboundary atmospheric haze pollution, were profound. The haze caused by the conflagration in the ASEAN region and elsewhere was directly linked to important issues of land use and abuse, toxic contamination, biodiversity conservation, greenhouse gas emissions, and particularly to the importance of fire management within an overall regime of land resource management.

Experience indicates that the underlying causes of fires cannot be fully removed and there is no easy remedy. Hence, abatement through effective and integrated fire management assumes great relevance and urgency.

Global Fire Occurrences

The annual rate of deforestation in developing countries during the 1980s was 16.3 million hectares (ha), while the corresponding figure for the developing countries of the Asian and Pacific region was 4.3 million ha. During the period 1990–1995, there was not much change in the rate of deforestation, standing at 13.7 million ha for all developing countries, and 4.2 million ha for forests of the Asian and Pacific region. While several causes have been attributed to the alarming rate of deforestation, in the majority of cases, fire has played a decisive role.

Every year, millions of hectares of the world's forests are being consumed by fires, big and small, resulting in billions of dollars in suppression costs and causing tremendous damage in lost timber, falls in real estate and recreational values, property losses, and deaths. Wildfire is influencing many aspects of our life: the flow of commodities on which we depend, the health and safety of the communities in which we live, and the health and resilience of wildland ecosystems.

There are many forests seldom affected by fire, while others regenerate easily after burning. Some forests are subjected to high fire frequencies and heavy destructive impact. It is difficult to estimate the number and extent of forest fires and related annual losses. Comprehensive reports on losses are not available and forest fire statistics are often deficient.

Prehistory of Fires

Wildfires have been present on the earth since the development of terrestrial vegetation and the evolution of the atmosphere. Lightning, sparks generated by swaying bamboos, and volcanoes have been some of nature's ways of igniting forests and keeping the plant environment dynamic. As a perfect relationship existed between fire and the ecosystem, such natural wildfires occurred at long intervals.

Taking a cue from nature, early humans used fire as a tool to alter their surroundings and later to prepare land for cultivation. There is paleontological evidence to show that fires occurred in the prehistoric past. The mythology of many countries features accounts of fires dating back to several thousands of years.

The climate in the tropical Amazonian region is too moist to allow a forest fire to burn as long as the forests are in an undisturbed state. There is, however, evidence that in the remote past, forest fires did occur in the region. For example, [experts have] determined that the ages of charcoal fragments collected from the Venezuelan part of the Amazon ranged from 250 to 6,260 years old. Those fires were most likely associated with extremely dry periods or human disturbances.

Recent History of Fires

Since the 1960s, several fires have attracted world attention. The Parana fire in Brazil in 1963 burned 2 million ha, destroyed more than 5,000 houses, and claimed 110 lives. With this started the new history of wildfires in Brazil, and a permanent worry, mainly with regard to the damage that fire can cause to forest plantations. The effect of fire on vegetation became an issue in 1988 following devastation to some parts of the Amazon forests. According to the World Wildlife Fund for Nature (WWF), large-scale logging and forest fires have contributed to the wiping out of some 12–15 percent of the Amazon rain forest. Satellite imagery from the Advanced Very High Resolution Radiometer (AVHRR) satellite, interpreted by the Brazilian National Institute for Space Studies, indicated that 20.5 million ha of Brazil's Legal Amazon (which covers an area of 500 million ha) was burned in 1987, of which about 8 million ha was considered to be deforestation in the dense forest area. In early 1998, the savannahs in the state of Roraima were left parched by the worst drought in history, resulting in big blazes, which burned some 3.2 million–3.5 million ha. Of this land, about 200,000 ha were good forests and the rest were already deforested areas or secondary forests.

Each year, fires in the Brazilian Amazon burn an area larger than Rio de Janeiro state. Ranchers and subsistence farmers ignite their lands in an attempt to convert forests into fields and to reclaim pastures from invading weeds. Subsistence farmers have migrated from all parts of Brazil to the fringes of the rain forest, lured by government promises of free land and a better life. But despite the lushness of the nearby forest, the soil is too poor to support intensive agriculture. They arrive armed with hope and chainsaws, clearing and burning land to farm the poor soil.

The fires set by ranchers and subsistence farmers often get out of hand, inadvertently burning forests, pastures, and plantations.

The results of a seven-year study conducted by the Woods Hole Research Center in Massachusetts, US suggest that the Amazon rain forest is experiencing an acute drying, leading to increased susceptibility to fire. Recent tests involving digging for water at many sites found dry ground, while similar tests seven years earlier revealed water close below the surface.

Statistics in the People's Republic of China (PRC) reveal that between 1950 and 1990, 4,137 people were killed in forest fires. In the same period, information from satellites reveals that about 14.5 million ha of forest were affected by fires in the neighboring Soviet Union, predominantly burning in the Siberian boreal forests, which have a composition similar to the northeastern PRC.

The Kalimantan fire in Indonesia in 1982 burned about 5 million ha and caused losses amounting to $9.1 billion. Fires have swept through the forests of Kalimantan and Sumatra (also elsewhere) in Indonesia several times during the last two decades, engulfing millions of hectares and causing losses estimated at billions of dollars.

In 1983, the Ash Wednesday fire in Australia caused 76 deaths, killed 300,000 sheep and cattle, and burned more than 2,500 homes. The Great Black Dragon Fire of the northern PRC in 1987 burned around 1.3 million ha, destroyed more than 10,000 houses, and resulted in a death toll of about 200. The Yellowstone fire in the United States in 1988 almost completely burned out one of the world's most famous parks, costing about $160 million to suppress, and causing an estimated loss of $60 million in tourist revenues between 1988 and 1990. In the longer term, however, the increased biodiversity created by the fires in Yellowstone National Park may well yield benefits that outweigh these losses.

In 1982–1983, Cote d'Ivoire in West Africa was swept by wildfires over a total area of about 12 million ha. The burning of some 40,000 ha of coffee plantations, 60,000 ha of cocoa plantations, and some 10,000 ha of other cultivated plantations had detrimental impacts on the local economy and left more than 100 people dead during this devastating fire period.

In the last four years, unusual weather conditions (and global weather changes) have led to fires in several parts of the world. Some of the conflagrations during 1996–1998 have been particularly damaging, as fires swept across the fragile rain forests of South America and two waves of forest fires gripped Indonesia, causing a national disaster.

Fires in Mexico and Central America have burned a reported 1.5 million ha. These have generated large quantities of smoke, which have blanketed the region and spread into the United States as far north as Chicago. From January to June 1998, about 13,000 fires burned in Mexico alone. Figures released by Mexican authorities in May 1998 indicated that a reduction of industrial production in Mexico City, which was imposed in order to mitigate the additional smog caused by forest fires, involved daily losses of $8 million.

Between December 1997 and April 1998, more than 13,000 fires burned in Nicaragua, the most in any Central American country, destroying vegetation on more than 800,000 ha of land. The Nicaraguan Ministry of Environment and Natural Resources recorded more than 11,000 fires in the month of April 1998 alone.

In July 1998, devastating forest fires affected more than 100,000 ha in eastern Russia. Coniferous forests burned in more than 150 locations around Vladivostok, Sakhalin, and Kamchatka peninsula. In Russia's Pacific island of Sakhalin alone, more than

25,000 ha of forest were consumed by fire during September 1998. The same year, forest fires in Florida, in the United States, burned an area of some 100,000 ha.

Fires burned the forests and pastures of Mongolia consecutively in each of the years between 1996 and 1998. The 1996 fires affected an area of 10.2 million ha, including 2.4 million ha of forests, in which 22 million cubic meters (m³) of forest growing stock were lost. The 1997 fires affected more than 12.4 million ha, of which forests accounted for 2.7 million ha. This fire killed some 600,000 livestock while damage to the Mongolian economy was estimated at $1.9 billion. . . .

There are predisposing factors or inherent conditions, as well as immediate causes that might result in wildfires and influence their frequency and intensity. These are often interrelated.

Predisposing Factors

The predisposing factors are various and may include the following:

- economic (poverty and dependence of rural communities on forests for livelihood);
- demographic (increased population pressure on forests for their goods and services);
- meteorological (weather conditions including high temperature and lower atmospheric humidity);
- crop conditions (amount of canopy opening causing desiccation and water stress, nature and amount of ground vegetation, and fuel load);
- nature and condition of ecosystem (vegetational types, fire resistance level of component species, and locational topography);
- sociocultural (cultural significance of fire to the forest dwelling and rural communities); and
- institutional (lax environmental laws, inadequate enforcement capability, indifference of public administration to environmental matters, lack of dissemination of weather information and fire danger warnings, misuse of funds earmarked for fire protection, and management and policy weaknesses).

Immediate Causes

The contribution of natural fires to tropical wildland fires today is negligible. Most tropical fires are set or spread accidentally or intentionally by humans, and are related to several causative agents, some of them linked to subsistence livelihood, others to commercial activities. These include:

- deforestation (conversion of forests to other land uses, e.g., as agricultural land and pastures);
- land clearance and land preparation for agricultural crops;
- traditional slash-and-burn agriculture;
- grazing land management (fires set by graziers, mainly in savannahs and open forests with distinct grass strata);
- extraction of nonwood forest products (NWFPs) (using fires to facilitate harvests or improve the yield of plants, fruits, and other forest products, such as honey, resin, and antlers, predominantly in deciduous and semideciduous forests);
- wildland/residential area interface fires (fires from settlements, e.g., from cooking, torches, camp fires, etc.);
- other traditional fire uses (in the wake of religious, ethnic, and folk traditions; tribal warfare);
- socioeconomic and political conflicts over questions of land property and land-use rights, using arson;
- speculative burning to stake land claims; accidental fires (e.g., due to falling of dry leaves and twigs on high tension electricity lines); and
- fires introduced by design (e.g., prescribed fires) going out of control and becoming wildfires.

Another factor to be noted in this regard is the connection between population growth and deforestation. The 1995 world population stood at 5.7 billion, and is expected to grow to about 9.4 billion by 2050, with all the attendant impacts on natural resources.

How to obtain a respite from deforestation and forest fires and haze is a major management challenge. . . .

Why People Burn the Land

Fire in the tropical rain forests is often related to forest clearing for agriculture, industrial timber plantations, and other land-use changes, of which three broad types can be distinguished.

- Shifting cultivation where land is allowed to return to forest vegetation after a relatively short period of agricultural use. Traditionally, shifting cultivation provided a sustainable base of subsistence for indigenous forest inhabitants and had little impact on forest ecosystem stability. Today, shifting cultivation is practiced by some 500 million people on a land area of 300 million–500 million ha and is often unsustainable due to the increase in size of individual plots and shorter fallow periods.

- Temporary complete removal of forest cover in preparation for industrial timber plantations.
- Permanent conversion of forests to grazing or cropland, as well as other nonforestry land uses.

In all cases, clearing and burning initially follow the same pattern; trees are felled at the end of the wet season and the slash is left to dry for some time. The efficiency of the first burning is variable; often not exceeding 10–30 percent of the ground biomass due to the large amount of forest biomass in tree trunks and stumps. The remainder is tackled by a second fire or left to decompose.

The burning of primary or secondary rain forest vegetation for conversion purposes has accelerated in recent years. Such forest-clearing fires often escape and have been shown to often lead to large-scale wildfires in undisturbed rain forests under the right climatic conditions.

While a land fire (e.g., in farmlands) may lead to a forest fire, a distinction is often made between the two in view of the differences in their causes, impacts, control measures, etc. Forest fires and associated haze have damaging impacts, but not designed and controlled burnings.

Some ecologists assert that fire is nothing more than a secondary factor in the destruction of dense and moist rain forests, which will not burn unless interference causes inflammable materials to be present or accumulated. Logging can increase the susceptibility of tropical rain forest to fire, particularly when carried out in an unnecessarily destructive and wasteful manner or resulting in large gaps in the forest canopy. Such practices can cause the accumulation of flammable biomass, invasion by weed species, and desiccation of soil organic matter—all factors that make forests susceptible to wildfire.

A series of disturbances may also increase the susceptibility of rain forests to fire. For instance, the extended rain forest fires of 1989 in Yucatan (Mexico) that burned some 90,000 ha were the result of a chain of disturbances. In 1987, Hurricane Gilbert damaged and opened the closed forests, leaving behind unusual amounts of downed woody fuels. These fuels were then desiccated by the drought of 1988–1989, and the whole forest area was finally ignited by escaped land-clearing fires. None of these three factors, the cyclonic storm, the drought, or the ignition source, if occurring alone, would have caused a disturbance of such severity.

Climate—An Aggravating Factor

Climate is a crucial control factor in fire occurrence and frequency, since it determines not only the vegetation, but also influences soil microorganism activity, and thus litter decomposition. In tropical lowland environments, litter decomposition is generally fast, and organic matter accumulation is rarely an important factor. However, climatic seasonality in terms of wetness and dryness is the most important parameter related to fire occurrence. Thus, a climate and vegetation analysis is imperative for an assessment of fire occurrence and frequency. Climatic seasonality does not manifest itself only on a month-to-month basis, but also in year-to-year variation.

If precipitation falls below 100 millimeters (mm) per month, and periods of two or more weeks without rain occur, the forest vegetation sheds leaves progressively. In addition, the moisture content of the surface fuels is lowered, while the downed woody material and loosely packed leaf-litter layer contribute to the buildup and spread of surface fires.

Aerial fuels such as desiccated climbers and lianas become fire ladders potentially resulting in crown fires or "torching" of single trees.

The occurrence of seasonal dry periods in the tropics increases with distance from the humid equatorial zone, leading to more open, semideciduous, and deciduous forest formations. Such forests are subject to frequent fires (often annual, but sometimes two or three times a year), and fire-tolerant species tend to dominate. The main fire-related characteristics of these formations are seasonally available flammable fuels (grass-herb layer, shed leaves). The most important adaptive traits that characterize the vegetation include thick bark, the ability to heal fire scars, resprouting ability, and seeds that feature fire adaptations.

Weather Variability

Meteorologists, based on available thermometric record, have determined that four of the hottest years in history occurred in the

1990s—1990, 1995, 1997, and 1998. The first five months of 1998 were the planet's hottest on record according to the scientists of the U.S. NOAA. The *El Niño* phenomenon is considered to be the main reason behind the mercury ascent and is frequently blamed for major forest fires. About 93 percent of all droughts in Indonesia have occurred during an *El Niño*. *El Niño* affects the global weather pattern, resulting in extreme dry conditions, which in turn leave forests parched and open to fires. Thus while *El Niño* is not a source of fire, it aggravates the danger of fires in places where negligence and management lapses can lead to severe conflagrations. Some point out that *El Niño* has always been in existence, without frequently causing major forest fires in the past. This may be due to the existence of relatively undisturbed forest cover with dense canopies in most tropical ecosystems that prevented drying of the lower vegetative strata, particularly the ground cover. . . .

Release of Greenhouse Gases

Forest fires contribute to global climate change and warming. Burning of forests also destroys an important sink for atmospheric carbon. Biomass burning is recognized as a significant global source of emissions contributing as much as 10 percent of the gross carbon dioxide and 38 percent of tropospheric ozone.

Depending upon the severity of annual burning and the weather, the emission products added to the atmosphere from biomass burning amount to 220–13,500 gigatons (Tg) of carbon dioxide, 120–680 Tg of carbon monoxide, 2–21 Tg of nitrous oxides, and 11–53 Tg of methane gas. The situation, obviously, is not very comforting. According to one estimate, in just a few months the burning that took place in 1997 in Indonesia released as much greenhouse gases as all the cars and power plants in Europe emit in an entire year. Scientists have estimated that from 1850 to 1980, between 90 billion and 120 billion metric tons (mt) of carbon dioxide were released into the atmosphere from tropical forest fires.

In comparison, during that same time period, an estimated 165 billion mt of carbon dioxide were added to the atmosphere by industrial nations through the burning of coal, oil, and gas.

According to recent estimates, some 1.8 billion–4.7 billion mt of carbon stored in vegetation may be released annually by wildland fires and other biomass burning. However, not all of the biomass burned represents a net source of carbon in the atmosphere. The net flux of carbon into the atmosphere is due to deforestation (forest conversion with and without the use of fire) and has been estimated to be in the range of 1.1 billion–3.6 billion mt per year.

Wildfires burning in radioactively contaminated vegetation lead to uncontrollable redistribution of radionuclides, e.g., the long-living radionuclides caesium (^{137}Cs), strontium (^{90}Sr), and plutonium (^{239}Pu). In the most contaminated regions of Belarus, Russian Federation, and Ukraine, the prevailing forests are young and middle-aged pine and pine-hardwood stands with high fire danger classes. In 1992, severe wildfires burned in the Gome region (Belarus), and spread into the zone of 30 kilometers (km) radius around the Chernobyl Power Plant. Research reveals that in 1990 most of the ^{137}Cs radionuclides were concentrated in the forest litter and upper mineral layer of the soil. In the fires of 1992, these radionuclides were lifted into the atmosphere. Within the 30-km zone, the level of radioactive caesium in aerosols increased 10 times. . . .

Social Impacts

Apart from causing transboundary air pollution, smoke emissions from wildfires affect human health, particularly causing respiratory ailments, and in some cases, disease and death. They also cause visibility problems, which may result in a breakdown of communication systems, accidents, and economic loss. Other social impacts include damage to energy and electric installations, disruption in the supply and distribution of food, displacement of people from affected areas, and temporary closure of educational establishments and production units.

Air Quality and Health

The main constituent of the smog that adversely affects health is particulate matter. The "WHO [World Health Organization] Health Guidelines for Episodic Vegetation Fire Events" that arose from the meeting of experts held in Lima, Peru, 6–9 October 1998, stressed the necessity of ground-based air quality monitoring of particulate matter in all countries affected by regional haze from vegetation fires. Ideally, PM_{25} (particulate matter with an aerodynamic diameter less than 2.5 microns) should be measured since that size fraction has a significant health impact. If that is not possible, PM_{10} or total suspended particulates (TSP) should be measured. The WHO draft document also recommends that additional pollutants such as ozone, nitric oxide, sulfur dioxide, carbon monoxide, aldehydes, and polyaromatic hydrocarbons be measured, if possible, to provide a comprehensive assessment of the health risks resulting from exposure to haze components. . . .

As the concentration of airborne particles rises, a corresponding rise in hospital admissions (and even deaths) is noted. Even though evidence is piling up increasingly on the lethal effects of particles, scientists are yet to fathom the extent of damage they can cause to public health.

Economic Impacts

Fire is one of the least expensive and simplest tools for preparing land for growing agricultural crops. However, once out of control, it can lead to long-term site degradation and other detrimental impacts. Fire can thus be a source of positive and negative impacts—the reason why rural societies considered fire as a good servant, but a bad master.

Apart from loss of material goods and services, forest fires cause serious direct economic losses through damage and decline in the quality of forest growing stock, reduced landscape stability, increased proneness to pests and diseases, reduced availability of forest-based raw material supplies, and the need for new investments in forest rehabilitation and fire protection. Indirectly, they affect agricultural productivity and tourism, indigenous populations and their means of livelihood, and jeopardize the prospects and ability of the rural poor to improve their standard of living.

Forest fires can also degrade other surviving forests by exerting impact on their composition, regeneration, productivity, protection functions, soil quality, wildlife, and aesthetics.

Controlled use of fire has been an important aspect of land and forest management. . . .

Since the beginning of the industrial revolution, humans have transformed about 40 percent of the earth's land surface. By 1990, conversion of tropical forests—much of it accomplished by open burning—had reached an estimated rate of 1.8 percent of the earth's total forestland per year. In absolute terms, this amounts to about 14.2 million ha. The implications of open burning on this scale for the global environment are startling.

A healthy forest is one that is resilient to changes. The term ecosystem health is used to define the structural and functional stability of an ecosystem and its ability to bounce back after stress. Forest fire management is an important aspect of sustainable forest management, ensuring the health of forest ecosystems, and that negative impacts of fire are minimized and positive impacts maximized. . . .

Fire Science and Technology

Fire research and technology development has received considerable stimulation from scientific projects conducted under the umbrella of the International Geosphere-Biosphere Program (IGBP) and other programs devoted to global change research. . . .

The application of technologies and methods of information gathering, processing, and distribution has revealed that many existing systems must be further developed to meet the requirements of precise and real-time application for early warning and management of fire and other environmental hazards.

IGBP at the Max Planck Institute for Chemistry, Freiburg, Germany, provides the basis for interdisciplinary fire research programs. One of the operational IGBP core projects is the International Global Atmospheric Chemistry (IGAC) Project, which is investigating the impact of biomass burning on the atmosphere and biosphere (BIBEX [Biomass Burning Experiment]). Since 1990, interdisciplinary international research campaigns have been conducted or are in the planning and implementation stage, the most important of which in the tropics are the Southern Tropical Atlantic Regional Experiment (STARE) and the Southeast Asian Fire Experiment (SEAFIRE).

STARE was designed to investigate the atmospheric chemical consequences of fires in tropical and subtropical forests and savannahs of South America (Brazil) and Southern Africa. This first intercontinental fire experiment was conducted in the field during 1992 and involved more than 150 fire researchers from 14 nations. It demonstrated that fires on both sides of the tropical Atlantic cause elevated ozone concentrations in the troposphere during the dry season (August–November). . . .

SEAFIRE will investigate the characteristics and regional and global transport of emissions from various types of fire in tropical Southeast Asia, such as fires used in forest conversion and shifting cultivation, and in grassland and seasonally dry monsoon forests.

How Wildfires Ignite and Burn

BY MARGARET FULLER

In this excerpt from her book Forest Fires: An Introduction to Wildland Fire Behavior, Management, Firefighting, and Prevention, *wilderness expert Margaret Fuller clearly presents the complex issues of how wildfires begin and grow. All fires, she explains, need fuel, heat, and oxygen—the fire triangle—to burn. Forests, especially in summer and during droughts, have each of these elements in abundance, which is why forest fires are so common. Fuller describes how lightning and people start most wildfires, and how weather, vegetation, and terrain determine what happens after the fire starts. Wildfires can either die out shortly after ignition, smolder for months unnoticed by people, or grow quickly into large blazes. Using the Yellowstone fires of 1988 as an example, Fuller describes the violent behavior of massive fires or conflagrations, which display phenomena like convection columns, firewhirls, and spotting, when embers jump ahead of a fire, starting new blazes. Margaret Fuller is a freelance writer, naturalist, wilderness guide, and instructor of mountain ecology and wilderness skills.*

W hen my son Stuart and I hiked down to the green oval of First Lieutenant Lake in Idaho's Salmon River Mountains, we smelled smoke, but when we reached the shore no one was there. At the first campsite, a small fire was burning in decayed wood and pine needles several feet from a fire ring. We yelled, "Hey, anyone here?" but no one answered, nor was anyone fishing on the gray ledges across the lake.

So we filled our canteens and pots in the lake and carried them back 50 yards to dump them on the foot-high flames. Two hours later the fire had cooled enough to leave. That evening we poured more water on nearby areas to douse any possible underground fire and camped on the other side of the lake next to the ledges. Luckily the flames were not any higher, or we could not have put the fire out without help.

If you have seen a similar fire, you may wonder how fire can move invisibly from a fire ring and come up somewhere else. How can the decayed wood catch fire so easily, when on rainy days you find it so difficult to start a camp fire?

Ignition and Heat Transfer

Knowing how something can catch fire may answer some of these questions. When burning cellulose (found in plants) and other carbohydrates react chemically with oxygen. This reaction, called *combustion*, produces carbon dioxide, water vapor, and heat. That is, in combustion, oxygen and fuel combine in the presence of heat. Combustion may or may not continue after the heat source is removed, and it can also occur without any external source of heat if the material can produce its own heat. Paint-soaked rags igniting in a closet are an example of this type of ignition called *spontaneous combustion*.

How do such fuels catch fire in the first place? To begin with, they must heat to a certain temperature (called the *ignition point*) in the presence of oxygen. Three things—fuel, heat, and oxygen—make up the *fire triangle;* if you remove one, the fire cannot burn. For example, if something cooking on a stove catches fire, pouring salt on it will put it out because it removes the air.

While the fuel heats to the ignition point, two things happen: First, the fuel dries out, because the heat boils away the water and drives off the volatile substances, like the resin in the pitch of pine trees. Second, the heat breaks down the chemical structure of the fuel, producing flammable gases, droplets of liquids called *tars*, charcoal, water, and ash. The amounts vary according to the temperature, with higher temperatures required to produce flammable gases. To ignite, or burst into flame, the wood must become hot enough to produce combustible gases on its surface and must produce enough of these gases to keep a flame burning. Wood will only glow or smolder if heated to 400° to 700° F, but at about 800° F it will burst into flame. Note that even though the flame of a match burns at 2300° F, it is so small that it can heat up only a small amount of dry fuel. . . .

When lightning or some other source ignites a forest or other wildland vegetation, three types of fire can result: surface, ground, and crown, often in combination. *Surface* fires move over the

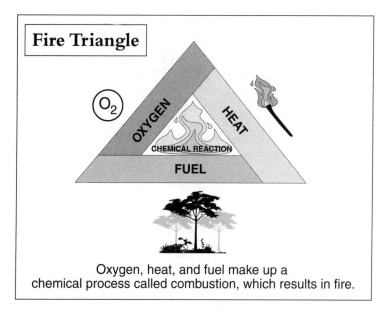

Fire Triangle

O₂
OXYGEN
HEAT
CHEMICAL REACTION
FUEL

Oxygen, heat, and fuel make up a
chemical process called combustion, which results in fire.

ground burning the litter on the surface, such as grasses, other plants, called forbs, and shrubs. They also burn dead branches and logs and small trees. The flames in a surface fire burn in a band called a *flaming front*, which is followed by glowing combustion. A surface fire also may "torch," or burn, all of a few isolated trees. *Ground* fires burn the duff and organic material in the soil beneath the surface litter of needles or leaves, usually by glowing combustion only. *Crown* fires burn the crowns or tops of trees or shrubs.

Fire Spread

A fire spreads horizontally by igniting a series of particles of fuel at or near its edge. At first the flames burn at one point, the source of ignition, and then move outward, accumulating enough heat to keep burning on their own. As the flames move out from the point of ignition, either wind or slope elongates the perimeter of the flames into an ellipse with a burned-out center. (Note that this ellipse is a prototype; not all fires burn in this way.) The *head* of the fire is the end of the ellipse that the wind blows ahead into new fuels. The head of the fire advances faster than do its sides or flanks because the leading flames are first to reach the unburned fuel preheating the fuel and drying it. Conversely, the rear of the fire is slanted toward the already-burned center, and so the flames move outward more slowly. As the speed of the wind increases, the ellipse becomes more elongated.

Slope affects a fire in a way similar to that of wind but, instead, stretches it uphill. Fire can also spread downhill or against the wind, but such a fire will be the flanks or rear of the ellipse formed by the main fire. Hence a *flanking* fire advances crosswind or across a slope, and a *backing* fire advances against the wind or downslope. In any case, the shape of a fire seldom remains elliptical, because barriers, slope, changes in fuel, and spotting cause it to develop fingers and even multiple heads. A fire with an expanding perimeter is a *line* fire.

In timber, crown fires can spread 5 or more miles per hour but in grass, fires spread at a rate of only 2 to 4 miles an hour. In forests in the West, fires usually advance less than one-quarter mile an hour. Perhaps surprisingly, high winds, like those in Yellowstone in 1988, may push along a fire too fast to permit it to do as much damage as it would if it moved more slowly. But the fuels—plants, shrubs, and trees—left behind may still burn after the flaming front has passed.

Fire Behavior

As seen in Yellowstone, prolonged drought lowers the moisture content of the fuel to the point that fire may spread quickly and burn with enormous flames. Park officials described the "behavior" of the Yellowstone fires as extreme. By *fire behavior* they meant the ways that the fuels ignited, the flames developed, and the fire spread. [Fire behavior analyst] Greg Zschaechner observed that typical fire behavior is controlled by the environment but that extreme fire behavior controls the environment.

Scientists studying the behavior of the Yellowstone fires in the fire behavior unit at the Intermountain Research Station's Fire Sciences Laboratory in Missoula, Montana, are constructing maps which show how far the fires spread each day. When drawing these maps they use images photographed from the planes that flew when the fires were burning, using special infrared cameras. When the planes were not able to fly into some areas because the fires were so large, the researchers gathered information from weather and fire records and fire managers.

First the lab associates digitized (put onto the computer) the tentative maps and then overlaid them with computerized terrain and fuels maps to verify the information. In a few places where they still could not determine the daily perimeters, they overlaid other fire-severity maps of Yellowstone made from satellite images. When they finish their maps, they will be able to analyze the fires' behavior by comparing their daily advances with other data. . . .

What influences the behavior of fires like those at Yellowstone? The three main influences are the particular fuels being burned, the weather, and the topography. Among these, fuel moisture, wind speed and direction, and slope steepness seem to have the greatest effect. Fuels with low moisture levels cause fires in fine dead fuels like grass to spread rapidly. Fuels with a high content of flammable oil, resin, or wax burn faster and at a lower moisture content. Wind speed is also important because strong winds can carry embers farther ahead of the flaming front, causing spot fires that accelerate the rate of spread. Spotting produced the fastest spread rates in the Yellowstone fires.

Topography

Topography is the physical shape and features of a region, and it affects fires by varying the weather within just a small area. . . . Because warm air rises, preheating uphill fuels, fires advance uphill faster than they travel downhill. A slope raises the fuels in front of the fire, thus bringing them closer to the flames, and also acts like a chimney carrying heat and flames uphill. Thus, depending on slope angle and wind speed, slope can be more important than wind in determining the rate of a fire's spread. . . .

Other topographical features, such as canyons, ridges, and bare areas, also influence fires. For example, a fire starting at the bottom of a slope is more likely to become large because more fuels are situated above it, and—remember—fire burns more easily uphill. Fires in narrow canyons preheat fuels across the canyon from them and also send embers across, as happened at Lowman [1989 Idaho fire]. A steep, narrow canyon pulls up heat and flames as in a chimney. . . .

Fire Intensity

Another characteristic of a fire is its *intensity*, or the rate at which a fire releases heat. . . .

The intensity of a fire is also related to the length of its flames, which limits the possible methods of suppressing it. When the flames are no more than 4 feet in length firefighters can usually attack the fire with hand tools. But when the flames are over 8 feet long, no control of any kind is likely to be effective. And when the flames are longer than 11 feet, fires often crown, spot, and make runs. (A *run* is rapid spread through surface or crown fuels.)

Crown Fires and Conflagrations

A *crown* fire is one that attacks the crown, or head of foliage, of a tree or shrub. . . .

To climb into the crowns of the trees, a fire first needs a *ladder*, in the form of a small tree or shrub or the dead lower branches of a larger tree. A crown fire also needs thick fuels and tree crowns in close proximity. Low moisture in fine fuels and low moisture and flammable chemicals in foliage also make crowning more likely. For a crown fire to stay in the tree crowns and keep moving, strong winds, steep slope, or high fire intensity must be present.

Because crown fires usually travel rapidly, they do not stay in one place long and so may be less destructive than the length of their flames would indicate. But when crown fires become very large and intense and spread quickly, fire managers rename them *conflagrations*, which are large destructive fires with moving fronts and rapid rates of spread. This type of fire occurred in Yellowstone in 1988. The extreme fire behavior of a conflagration differs from a firestorm because it moves faster. In addition, crown fires, conflagrations, and fire storms often throw embers a long distance ahead of themselves. This phenomenon is called *spotting.*

Spotting

Spotting occurs when the hot air rising from the fire carries embers or firebrands (pieces of burning wood) upward and over to stands of trees ahead of the fire, where they start new fires. Long-range spotting requires firebrands large enough to travel long distances and to stay burning until they hit the ground. When the temperature is high and the humidity and fuel moisture are low, almost all the firebrands can start fires. In Yellowstone, spotting embers flew as much as 1½ miles ahead of the fires, but even 5 miles is not uncommon. In Tasmania in 1983, one ember spotted 47 miles from the source, according to Tony Mount, a fire research specialist. In this fire, curled strips of eucalyptus bark that were burning inside were transformed into flaming javelins by 100-mile-an-hour winds.

Smoke is the combination of gases, water vapor, soot, and

other hydrocarbons distilled by the heat of the fire. Smoke rises because heated air rises. A small fire can burn with a pale, almost transparent smoke plume that shows no real motion, or it can burn intensely, giving off a dense plume of smoke with definite edges that expand as it rises. Fire managers call this dense plume a *convection column* because it is formed by convection, or the motion of hot air. Although moderate winds may tilt the column, stronger winds prevent a column from forming at all.

When a convection column becomes large, so much warm air rises that it pulls the cool air around the fire into it, creating a noticeable indraft into the fire and giving birth to a veritable whirlwind of fire.

Jack Wilson, the director of the Bureau of Land Management at BIFC [Boise Interagency Fire Center], counted one evening in Yellowstone in 1988 five convection columns at or above the 35,000-foot altitude of the plane he was in. Around sunset one evening when flying over a ridge at the head of Slough Creek, Doug Brown, a helitack crewman for the Sawtooth National Forest, saw the evening winds pushing fire with 300-foot flames and huge convection columns down the canyons. Indeed, the heat and winds caused by these columns' sucking in air forced him and his pilot to leave the area, but when they tried to leave, the 60-mile-an-hour winds at first held the helicopter at a standstill. Later, when the fire forced them to evacuate the helibase for the main Storm Creek Fire camp, winds from the fire tossed around medium-sized helicopters like chunks of bark. As the wind blew the tail of the helicopter in which Brown was riding at a 30 degree angle, the helicopter kept dropping, 50 to 100 feet at a time.

When conditions in the fire environment are severe, sometimes nothing can control the wildfires that result, even if firefighters reach them right after they start. As Frank Carroll of the Boise National Forest observed, "We don't manage large wildfires; they do whatever the hell they please."

Firestorms

We heard a lot about firestorms in the news reports from Yellowstone in 1988, even though those fires were really conflagrations. What is the difference? Fire managers define a *firestorm* or a mass fire as a large area of intense heat that causes violent convection. This definition also applies to heavy aerial bombing, such as the firestorm created in Hamburg, Germany, during World

War II. According to fire behavior expert Richard Rothermel, a firestorm is a very high intensity event with a large convection column that throws off embers in a shower of short-range spotting all around itself.

Such a fire acts as though a whole area of trees, rather than just one tree, were torching, or it may appear that the flame front has suddenly grown much wider. Whether or not a fire is a firestorm, however, depends on its actions rather than its size. Firestorms often include destructive, violent surface indrafts and sometimes contain tornadolike whirls. In fact, they may act as though several large whirlwinds or convection columns have combined.

What causes firestorms? In a study of blow-up fires in the South, George Byram concluded that the principal cause is a buildup of heat in one area, but fire scientists still know little else about firestorms. They believe that a wind speed decreasing with height above the fire, unstable air, and the advance of the fire into heavy fuels may add to the buildup of heat. This in turn strengthens the winds whirling up the convection column enabling them to overcome any horizontal winds. A firestorm can occur without the prerequisite of a large fire. As seen at the Lowman Fire, in the right weather, a fire of moderate size spotting into an area of heavy fuels can cause a firestorm.

Firestorms are unpredictable because they can develop suddenly when large wind-driven fires slow down. Jack Lyon, a research biologist at the Intermountain Research Station Forestry and Sciences Laboratory in Missoula, Montana, studied a firestorm that arose along the Pack River in the 1967 Sundance Fire in the Selkirk Mountains of northern Idaho. Lyon's study combined observation of the fire effects, official records, and eyewitness accounts. The firestorm was created when the fire ran 16 miles in 9 hours as it burned 50,000 acres. A prolonged dry spell, high temperatures, a 4-mile-long fire front, unstable air, and a dry cold front combined to cause the run, which was considered a conflagration. It then became a firestorm when it stalled in the Pack River Valley. During the storm, 80-mile-an-hour winds snapped off trees 60 to 75 feet above the ground. Lyon calculated the peak fire intensity as equal to a 2000-kiloton nuclear bomb. . . .

Fire Whirls

A spinning, moving column of air, called a *fire whirl*, carries flames, smoke, and debris aloft and forms a *vortex* (spinning air that forms a vacuum in the center). The vacuum draws in flames and debris. Such fire whirls can range in size from dust devils to tornadoes. Small fire whirls do not have much effect on fire, but larger whirls cause spotting and increase the fire's intensity, because when a fire whirl forms, the combustion rate inside it increases dramatically.

Residential Fire Facts

BY FRANK FIELD AND JOHN MORSE

Fires kill more Americans every year than all other natural disasters combined, including floods, hurricanes, tornadoes, and earthquakes. Seventy-four percent of fire-related deaths and 66 percent of fire injuries occur in private homes.

In this excerpt from their book Dr. Frank Field's Get Out Alive: Save Your Family's Life with Fire Survival Techniques, *Dr. Frank Field and John Morse describe where and how residential fires occur. According to Field and Morse, the kitchen and chimneys are the places where most home fires start. The authors indicate that smoking, arson, cooking, and heating are the most common sources of ignition. Field and Morse based their book on the Emmy Award–winning CBS News series* Plan to Get Out Alive, *which was sponsored in part by the U.S. Fire Administration in an effort to prevent senseless fire tragedies. Dr. Frank Field works for CBS News and is among the most popular and respected of news personalities. John Morse is an artist and writer living in New York City.*

A lthough the damage and havoc fire can wreak are clear, the causes of fire are much less obvious. In learning how to prevent fires, it's important to know why and where fires strike. Seventy-four percent of fire deaths and 66 percent of fire injuries occur in private residences, but while hotel and office fires grab the headlines, fire deaths at home often go unnoticed by the media.

Causes of Fire

Fires are measured in several ways—by fire deaths, fire injuries, dollar loss, and fire incidence. For each of these four categories, there is a different set of causes. The three leading causes of fire deaths in private homes are careless smoking, arson, and heating. However, of the three leading causes of residential fires overall, heating ranks number one, cooking accidents is second, and arson ranks third.

Careless Smoking

Misuse of smoking materials is the leading cause of all residential fire deaths, even though it does not rank as one of the top three causes of residential fires overall. Although many deaths result from smoking in bed, more often, fires that cause death are started as a result of smoking in the lounging areas of the home, such as the living room, den, or playroom. Typically a cigarette falls between the furniture cushions and upholstery—often dropped by someone who has been drinking—where it smolders for hours before bursting into flames. This lapse of time between dropping the cigarette and the actual start of the fire makes it easy to see how a furniture fire can catch victims unawares and lead to death.

In 1991 three Brooklyn teenagers perished in a fire caused by the careless smoking habits of a downstairs neighbor. Despite the deployment of a team of 90 firefighters, the lives of Tony, Penelope, and Donnel were ended by a burning cigarette dropped into a mattress on the floor below their apartment.

> That same year English firefighters couldn't save Steve Marriott, the lead musician of the rock group Humble Pie, best known for his recording of "Itchycoo Park." In April 1991 he had just returned to his 16th-century cottage in Essex after a two-week recording session. Marriott had high hopes for a comeback, but his career came to a brutal end in a fire, probably caused by a cigarette left unattended in his bedroom.

Arson

"Arson" is a legal term with a somewhat narrow definition. It's also a term used to describe the broader fire category known as "incendiary and suspicious." Under any name it's deadly.

> The 1991 fire at the Happy Land Social Club in New York City was traced to arson. Eighty-seven people died in a matter of minutes, leaving scores of children

orphaned and hundreds of lives forever changed, all because one man was seeking revenge against a woman
who had rebuffed him.

Julio Gonzalez was later convicted of purposely setting
the fire to get back at his ex-girlfriend. Gonzalez told
how, after an argument with the woman at the club, he
filled a one-gallon plastic container with a dollar's
worth of gasoline. He returned to the club and doused
the entrance of the crowded dance hall, lit a match, ignited the door, then went home to sleep.

Whether Gonzalez intended to kill everyone isn't important (the ex-girlfriend was one of only a handful of
survivors). Scores of people were murdered that night.

Gonzalez's intentions remain murky, but the disgruntled workers who started the San Juan Dupont Plaza
Hotel fire later told the FBI and fire inspectors that
they hadn't meant to kill anybody, just to cause trouble
for their employer. Nevertheless, 97 people died and
scores more were injured.

Arson is the third leading cause of residential fires, and it is on
the increase. Arson is the number one cause of fire dollar loss.
Fraud—usually bilking an insurance company—remains the leading motivation for arson, but many other reasons have been cited,
including vandalism, revenge, and quarrels. Fire has become an
increasingly popular method for expressing anger.

While major disasters like the Happy Land Social Club and
Dupont Plaza Hotel infernos get the attention of the public and
underscore the tragedy of arson, it's important to remember that
most people who die in arson fires are killed in groups of one,
two, or three, and that these crimes are committed every day,
every year.

Cooking

Cooking is the number one cause of injuries and the second
leading cause of all residential fire. Most kitchen fires can be
traced to heat sources that are left unattended while cooking.
These accidents could easily be prevented by being more vigi-
lant during meal preparation. In addition, knowing how to use

A house fire burns out of control. Nearly one-third of all residential fires start in the kitchen.

baking soda or a pot lid to extinguish a grease fire (like knowing how to stop, drop, and roll when your clothes catch fire) would help reduce the injury rate associated with cooking fires.

Twenty years ago, cooking fires were the leading cause of all residential fires, but as more and more people have turned to alternative heating systems such as wood stoves and space heaters since the energy crisis of the 1970s, heating has replaced cooking as the cause of most fires.

Heating

Nothing causes more residential fires than heating hazards. The incidence of home fires jumps dramatically during the winter months. Bad wiring, poor ventilation, and dirty exhaust ducts contribute to the problem, but mostly it's misuse of space heaters, fireplaces, central heating systems, and wood-burning stoves that make heating the nation's number one residential fire hazard.

> In New York City, a tragic 1989 fire that killed three children and critically injured their parents was traced to a faulty electrical space heater. A fourth child, 13-

year-old April, had been testing the heater in various outlets, trying to find one that wouldn't make it shoot sparks. After finding one in the living room, she plugged in the heater and went to bed.

Everyone in the house was asleep when the fire broke out. The two-story home had two smoke detectors, but batteries had been removed from both of them. April had closed her bedroom door, so she managed to escape unharmed. Her mother was admitted to the hospital in critical condition, suffering from severe smoke inhalation. The father received third-degree burns over 56 percent of his body. He had attempted to rescue his youngest child, but ended up throwing himself through a window on the second floor to escape the flames.

One firefighter told the The New York Times the next day that he had crawled on his stomach through the thick smoke and the intense heat, trying to save two of the children. "I was going just on the sense of touch. That's when I felt the little girl. She was lying on the floor. I reached up onto the bed next to her and felt the little boy. They were lifeless."

The most recent figures available [by 1992] from the U.S. Fire Administration reveal that one of five residential fires is related to heating. Additionally, one of ten fire deaths results from heating problems. Heating fires are the third leading cause of death and the second leading cause of dollar loss.

The U.S. Consumer Product Safety Commission estimates that in 1988, 140 Americans died as a result of portable heaters. The fires caused $43 million in property damage.

This critical component of the residential fire problem includes fireplaces, portable space heaters, wood stoves, water heaters, fixed room heaters, and central heating. Because centralized heating systems in apartment complexes are normally subject to regular professional maintenance, heating is the cause of only about seven percent of fires in this type of residence. Most fires occur in one- and two-family homes. Of all the non–central heating fires, 75 percent are caused by human error, such as faulty installation. Even the most safely designed unit can be a hazard.

The tragedies surrounding heating fires are rife with preventable mistakes. Their victims are of all ages, but the very young and the very old are at greatest risk. Preventing these horrors requires constant vigilance and extreme caution with any kind of heating system, particularly freestanding heaters.

Besides the burns that result from touching these devices, there is also the danger that nearby materials could catch on fire, such as furniture, walls, curtains, carpet, and clothing. And there is grave danger in the smoke and gases heaters often create. . . .

Chimneys

Chimneys are the lungs of a fuel-burning system. If they have leaks and cracks, or if they are too small, plugged up, or clogged with soot and creosote, they can shut down the respiratory system of the heater, just as an injury can collapse your lungs.

When you decide to use a fuel-burning heater in your home, you should first consider whether or not your chimney is up to the job. Just because a heating system provides you with lots of warmth, that doesn't mean that it won't rob the air of oxygen.

> Two teenage boys in Minnesota recently paid the ultimate price for not understanding the need for adequate ventilation. Even though they had opened a window in their tiny cabin, it wasn't enough to ventilate their wood-burning stove. While they slept soundly in their cozy, warm cabin, the stove used up all the oxygen in the air. Both boys suffocated.

Every year, millions of new fireplaces and stoves are installed in the United States. The popularity of this type of heating system has led to a parallel growth in carbon monoxide deaths due to improperly ventilated heating systems. Each year, about 400 Americans die from carbon monoxide poisoning due to faulty home heating systems.

Carbon monoxide is an insidious poison, a colorless, odorless gas that lulls its victims into a deep sleep that ends in death. It catches its victims unaware, leaving them disoriented and groggy. Survivors describe its effect as debilitating, something akin to being paralyzed. . . .

Chimneys that exhaust smoke to the outdoors should never be any smaller than the pipe leading from the stove or heater. The flue should always be large enough for the pipe. Variance leading to poor ventilation is a health threat.

Begin by inspecting the outside of your chimney visually. Do you notice any creosote bleeding through the mortar lining? (It will resemble dark streaks of paint.) Is the chimney deteriorating in any way? Is mortar breaking away from the bricks? Are the bricks between the chimney and the roofline loose? Are bricks cracked, especially at offsets, the places where the bricks on the next row meet? Is the brick mortar hard, or would a sharp object easily pierce it? Does flashing—protective metal plates—surround the chimney and separate it at least two inches from roof shingles? Do you see any blistering of paint at any point where the chimney is behind a wall?

Signs such as these indicate a tear in the chimney lining that could lead to a heat leak, exposing the roof and interior of the house to extremely high temperatures and posing a fire risk. The flue, which regulates the air between the burning fuel and the chimney, should also be cleaned regularly to prevent soot and creosote buildup. . . .

Electrical Heaters

Electrical space heaters offer a relatively safe heating alternative, but like any heating device, they must be handled with extreme caution. Every year, improperly handled electrical heaters cause deaths, injuries, and hundreds of millions of dollars in property losses.

Whenever using space heaters of any type, keep all flammable materials at least three feet away. Direct contact with the heating element can cause burns. Check the heater regularly for cracked or frayed wires. Make sure that there is an automatic turnoff switch in case the heater accidentally tips over. The automatic turnoff switch should also work when the unit overheats.

Watch for overloaded wiring when using electrical space heaters. Overloaded wiring causes fires by overheating cords.

When purchasing space heaters, always look for the UL (Underwriters Laboratories) symbol, a sign that the unit has been approved by an independent testing agency.

Children Playing with Fire

Children who play with matches are the source of one of the most tragic fire statistics. Playing with fire increased rapidly in the 1980s as a cause of fires and fire injuries; it now ranks as the third leading cause of fire injuries. Furthermore, children playing with fire constitutes the fourth leading cause of residential fire deaths.

Similarly, playing with open flames—a category that includes candles, matches, lighters, embers, ashes, sparks, and torches—represents the third leading cause of dollar loss in residential fires. This is a growing cause of fires in the category of residential fire deaths and injuries.

When Fires Occur

Fire is not random and neither is the time of day that it strikes. Over the years, emerging trends have shown that the majority of fires and fire deaths occur at distinct times of the days.

Most deaths caused by fire happen late at night. Over half of the residential fire deaths occur between 11 P.M. and 6 A.M. Most people who die in fires die in their sleep. The worst hour is between 2 and 3 A.M.

Fire injuries, on the other hand, happen throughout the day. The most likely time for a fire injury is between 7 and 8 P.M. Unlike fire deaths, injury levels actually drop in the early morning hours when people are asleep. The peak hours of fire injury correspond to the hours when most people are preparing dinner. Fire incidence, which peaks from 6 to 7 P.M. also corresponds to meal times. Although fire dollar loss reaches its high point at the same time most fire deaths occur, like fire injuries, it drops in the early morning hours before dawn.

Winter, as might be expected, is the season when a home is most likely to catch fire and also the time when residential fire deaths are most likely to occur. In fact, a residential fire death is almost three times as likely to occur in January as in September. Nearly one-fourth of the entire year's fire deaths occur during the months of December and January alone. Heating is the main cause of the deaths, but other factors, such as holiday-related activities and increased alcohol consumption, also influence these statistics.

Where Fires Occur

Knowing where in the home fires typically occur can help in preventing injury, death, and financial loss. The most likely area is the kitchen. Most fires are associated with cooking over a stove or in the oven, but secondary areas, such as the exhaust fan, can also be troublesome. The most common reason for fires in the kitchen is that food has been left unattended during cooking.

Chimneys are the second most likely location for a home fire. When chimneys are not cleaned on a regular basis, soot and creosote accumulate; this material itself can catch on fire.

The third most likely location is the bedroom. In this living area, there are several leading factors, including—but not limited to—smoking in bed, children playing with matches, and arson. A typical scenario features a fire started by a child hiding under a bed or in a dark closet and igniting a match or cigarette lighter.

Garage and storage areas round out the list of the top five locations where fires start. Gasoline stored in unsealed containers leaks fumes that can travel along the floor and ignite upon contact with a spark or flame. About 20,000 such fires were reported in 1987 in the United States.

Where Fire Deaths Occur

For residential fire deaths, the breakdown is completely different. The most likely place for a person to die from a residential fire is in one of the lounging areas of the home, such as the living room. Typically, a careless smoker who has been drinking becomes intoxicated and falls asleep in a chair or couch holding a lighted cigarette that drops into the furniture and bursts into flames hours later. Contrary to what most people think, a person is more likely to be killed from careless smoking in the living room than the bedroom.

Bedrooms are the second most likely place for fire deaths. Again, smoking is a leading cause, but arson also plays a significant role, as does children playing with fire.

The kitchen is the third most dangerous room, and cooking is the leading cause. . . .

Dining areas and multipurpose rooms complete the list; about two percent of all fire deaths occur there. For around seven percent of residential fire deaths, the place of the fire's origin is unknown.

Apartment Fires

Although the statistics cited above include residences of all types in the United States, they largely reflect one- and two-family dwellings, where three-quarters of the population lives. But many Americans—particularly those in urban areas—live in apartment buildings. These buildings, where numerous people from various backgrounds live under the same roof, have unique conditions that result in somewhat different causes for fires and fire deaths.

The most notable difference between one- and two-family

homes and apartment buildings involves heating-related fires. Because most apartment buildings have centralized heating systems, heating ranks far lower as a cause of apartment fires than for fires in other types of residences.

As in dwellings of all types, careless smoking is the number one cause of fire deaths in apartments. However, the second and third leading causes of apartment fire deaths are different. After smoking, arson is the most common reason, followed closely by children playing with fire. Surprisingly, these three causes—smoking, arson, and children playing—account for over two-thirds of fire deaths in apartments. All other causes account for a relatively small percentage of all apartment fires.

The causes of apartment fires are significantly different from causes for residences of all types. Heating, the single most likely source of fires in all residences, ranks number five as the cause of apartment fires. Cooking accounts for about one-third of all fires in apartments, followed by arson, smoking, and a category known as "other equipment," which includes maintenance equipment such as incinerators or elevators, computers, office machinery, and generators.

Apartment fire deaths follow the general profile for residences of all types: Fires that kill are most likely to occur in the early morning hours, while the most typical hour for a fire of any type to strike is during the dinner hours, due to the association with cooking.

One major difference is the lack of variance from month to month when fires occur. This reflects the fact that heating accounts for a smaller percentage of fires in apartments than in residences in general. Because many people stay home during the winter months (and smoke and drink) and because children may get bored and start playing with matches for fun, winter is still the peak season for incidences of fire, although less dramatically so than in dwellings of all types.

Famous Fire Disasters

The Great Fire of London

By Samuel Pepys

In 1666 a conflagration broke out in London, which eventually burned four-fifths of the city. The fire burned for five days, killing twenty people, destroying eighty-seven churches, and leveling thirteen thousand houses. After experiencing two decades of difficulties, including civil wars and a plague, England was especially ill prepared for this tragedy.

In the following selection, Londoner Samuel Pepys reports his observations, which he recorded in his diary, of the great fire. Each day he wrote about what he observed as people scrambled to save their possessions and government officials struggled to deal with the catastrophe. Pepys held numerous government positions during a long public career.

September 2 (Lord's Day), 1666

Some of our maids sitting up late last night to get things ready against our feast today, Jane called us up about three in the morning, to tell us of a great fire they saw in the City. So I rose and slipped on my nightgown,[1] and went to her window, and thought it to be on the back-side of Marke-lane at the farthest; but, being unused to such fires as followed, I thought it far enough off; and so went to bed again and to sleep. About seven rose again to dress myself, and there looked out at the window, and saw the fire not so much as it was and further off. By and by Jane comes and tells me that she hears that above 300 houses have been burned down tonight by the fire we saw, and that it is now burning down all Fish-street, by London Bridge. So I made myself ready presently, and walked to the Tower,[2] and

---

1. dressing gown   2. London Tower

there got up upon one of the high places, Sir J. Robinson's little son going up with me; and there I did see the houses at that end of the bridge all on fire, and an infinite great fire on this and the other side the end of the bridge; which, among other people, did trouble me for poor little Michell and our Sarah on the bridge. So down, with my heart full of trouble, to the Lieutenant of the Tower, who tells me that it begun [*sic*] this morning in the King's baker's house in Pudding-lane, and that it hath burned St. Magnus's Church and most part of Fish-street already. So I down to the water-side, and there got a boat and through bridge, and there saw a lamentable fire. Poor Michell's house, as far as the Old Swan, already burned that way, and the fire running further, that in a very little time it got as far as the Steele-yard, while I was there. Everybody endeavoring to remove their goods, and flinging into the river or bringing them into lighters[3] that lay off; poor people staying in their houses as long as till the very fire touched them, and then running into boats, or clambering from one pair of stairs by the water-side to another. And among other things, the poor pigeons, I perceive, were loth to leave their houses, but hovered about the windows and balconys till they were, some of them burned, their wings, and fell down. Having stayed, and in an hour's time seen the fire rage every way, and nobody, to my sight, endeavoring to quench it, but to remove their goods, and leave all to the fire, and having seen it get as far as the Steele-yard, and the wind mighty high and driving it into the City; and every thing, after so long a drought, proving combustible, even the very stones of churches, and among other things the poor steeple by which pretty Mrs.— lives, and whereof my old schoolfellow Elborough is parson, taken fire in the very top, and there burned till it fell down: I to White Hall[4] (with a gentleman with me who desired to go off from the Tower, to see the fire, in my boat); to White Hall, and there up to the King's closet in the Chapel, where people come about me, and I did give them an account dismayed them all, and word was carried in to the King. So I was called for, and did tell the King and Duke of York what I saw, and that unless his Majesty did command houses to be pulled down nothing could stop the fire. They seemed much troubled, and the King commanded me to go to my Lord Mayor from him, and command him to spare no houses, but to pull down be-

3. flat-bottomed barges   4. a palace in central London

fore the fire every way. At last met my Lord Mayor in Canning-
street, like a man spent, with a handkerchief about his neck. To
the King's message he cried, like a fainting woman, "Lord, what
can I do? I am spent: people will not obey me. I have been
pulling down houses; but the fire overtakes us faster than we can
do it." People all almost distracted, and no manner of means used
to quench the fire. The houses, too, so very thick thereabouts,
and full of matter of burning, as pitch and tar, in Thames-street;
and warehouses of oil, and wines, and brandy, and other things.
And to see the churches all filling with goods by people who
themselves should have been quietly there at this time. Met with
the King and Duke of York in their barge, and with them to
Queenhithe, and there called Sir Richard Browne to them. Their
order was only to pull down houses apace, and so below bridge
at the water-side; but little was or could be done, the fire coming upon them so fast. River full of lighters and boats taking in
goods, and good goods swimming in the water, and only I observed that hardly one lighter or boat in three that had the goods
of a house in, but there was a pair of Virginals[5] in it. So near the
fire as we could for smoke; and all over the Thames, with one's
face in the wind, you were almost burned with a shower of fire-
drops. This is very true; so as houses were burned by these drops
and flakes of fire, three or four, nay, five or six houses, one from
another. When we could endure no more upon the water, we to
a little ale-house on the Bankside, over against the Three Cranes,
and there stayed till it was dark almost, and saw the fire grow; and,
as it grew darker, appeared more and more, and in corners and
upon steeples, and between churches and houses, as far as we
could see up the hill of the City, in a most horrid malicious
bloody flame, not like the fine flame of an ordinary fire. Barbary
and her husband away before us. We stayed till, it being darkish,
we saw the fire as only one entire arch of fire from this to the
other side the bridge, and in a bow up the hill for an arch of
above a mile long: it made me weep to see it. The churches,
houses, and all on fire and flaming at once; and a horrid noise the
flames made, and the cracking of houses at their ruin. So home
with a sad heart, and there find every body discoursing and
lamenting the fire; and Poor Tom Hater come with some few of

5. a small, rectangular spinet without legs, usually spoken of in the plural as a pair of
virginals

his goods saved out of his house, which is burned upon Fish-street Hill. I invited him to lie at my house, and did receive his goods, but was deceived in his lying there, the news coming every moment of the growth of the fire; so as we were forced to begin to pack up our own goods, and prepare for their removal; and did by moonshine (it being brave dry, and moonshine, and warm weather) carry much of my goods into the garden, and Mr. Hater and I did remove my money and iron chests into my cellar, as thinking that the safest place. And got my bags of gold into my office, ready to carry away, and my chief papers of accounts also there, and my tallies into a box by themselves.

September 3, 1666

About four o'clock in the morning, my Lady Batten sent me a cart to carry away all my money, and plate, and best things, to Sir W. Rider's at Bednall-Greene. Which I did, riding myself in my nightgown in the cart; and, Lord! to see how the streets and the highways are crowded with people running and riding, and getting of carts at any rate to fetch away things. The Duke of York come this day by the office, and spoke to us, and did ride with his guard up and down the City to keep all quiet (he being now General, and having the care of all). At night lay down a little

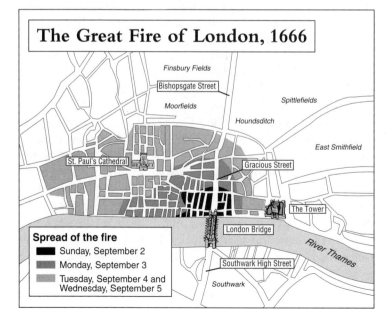

The Great Fire of London, 1666

upon a quilt of W. Hewer's in the office, all my own things being packed up or gone; and after me my poor wife did the like, we having fed upon the remains of yesterday's dinner, having no fire nor dishes, nor any opportunity of dressing anything.

September 4, 1666

Up by break of day to get away the remainder of my things. Sir W. Batten not knowing how to remove his wine, did dig a pit in the garden, and laid it in there; and I took the opportunity of laying all the papers of my office that I could not otherwise dispose of. And in the evening Sir W. Pen and I did dig another, and put our wine in it; and I my Parmazan cheese, as well as my wine and some other things. Only now and then walking into the garden, and saw how horridly the sky looks, all on a fire in the night, was enough to put us out of our wits; and, indeed, it was extremely dreadful, for it looks just as if it was at us, and the whole heaven on fire. I after supper walked in the dark down to Tower-street, and there saw it all on fire, at the Trinity House on that side, and the Dolphin Tavern on this side, which was very near us; and the fire with extraordinary vehemence. Now begins the practice of blowing up of houses in Tower-street, those next the Tower, which at first did frighten people more than anything; but it stopped the fire where it was done, it bringing down the houses to the ground in the same places they stood, and then it was easy to quench what little fire was in it, though it kindled nothing almost. Paul's[6] is burned, and all Cheap-side. I wrote to my father this night, but the post-house being burned, the letter could not go.

September 5, 1666

About two in the morning my wife calls me up and tells me of new cries of fire, it being come to Barking Church, which is the bottom of our lane. I up, and finding it so, resolved presently to take her away, and did, and took my gold, which was about £2350, W. Hewer, and Jane, down by Proundy's boat to Woolwich; but, Lord! what a sad sight it was by moonlight to see the whole City almost on fire, that you might see it plain at Woolwich, as if you were by it. There, when I come, I find the gates shut, but no guard kept at all, which troubled me, because of dis-

6. St. Paul's Cathedral

course now begun, that there is plot in it, and that the French had done it. I got the gates open, and to Mr. Shelden's, where I locked up my gold, and charged my wife and W. Hewer never to leave the room without one of them in it, night or day. So back again, by the way seeing my goods well in the lighters at Deptford, and watched well by people. Home, and whereas I expected to have seen our house on fire, it being now about seven o'clock, it was not. I up to the top of Barking steeple, and there saw the saddest sight of desolation that I ever saw; everywhere great fires, oil-cellars, and brimstone,[7] and other things burning. I became afraid to stay there long, and therefore down again as fast as I could, the fire being spread as far as I could see it; and to Sir W. Pen's, and there eat a piece of cold meat, having eaten nothing since Sunday, but the remains of Sunday's dinner.

September 6, 1666

It was pretty to see how hard the women did work in the can-nells, sweeping of water; but then they would scold for drink, and be as drunk as devils. I saw good butts of sugar broke open in the street, and people go and take handfuls out, and put into beer, and drink it. And now all being pretty well, I took boat, and over to Southwarke, and took boat on the other side the bridge, and so to Westminster, thinking to shift myself,[8] being all in dirt from top to bottom; but could not there find any place to buy a shirt or pair of gloves. A sad sight to see how the River looks; no houses nor church near it, to the Temple, where it stopped.

September 7, 1666

Up by five o'clock; and blessed be God! find all well; and by wa-ter to Paul's Wharf. Walked thence, and saw all the town burned, and a miserable sight of Paul's church, with all the roofs fallen, and the body of the quire fallen into St. Fayth's; Paul's school also, Ludgate, and Fleet-street, my father's house, and the church, and a good part of the Temple the like. This day our Merchants first met at Gresham College, which, by proclamation, is to be their Exchange. Strange to hear what is bid for houses all up and down here; a friend of Sir W. Rider's having £150 for what he used to let for £40 per annum. Much dispute where the Customhouse

7. sulfur   8. to change clothes

shall be; thereby the growth of the City again to be foreseen. I home late to Sir W. Pen's, who did give me a bed; but without curtains or hangings, all being down. So here I went the first time into a naked bed, only my drawers on; and did sleep pretty well: but still both sleeping and waking had a fear of fire in my heart, that I took little rest. People do all the world over cry out of the simplicity of my Lord Mayor in general, and more particularly in this business of the fire, laying it all upon him.

Chicago's Historic Blaze

By David Cowan

In the following selection, excerpted from his book Great Chicago Fires, *David Cowan explains how the fast, disorganized growth of Chicago turned it into an urban tinderbox. Cowan believes that the combination of hot, dry weather, a preponderance of wood buildings, and a scarcity of fire stations was responsible for the fire's destructiveness. When the fire, which began on October 8, 1871, finally burned itself out on October 9, more than fifteen thousand buildings had been destroyed and more than three hundred people had been killed. On a positive note, Cowan claims that the great fire, by leveling most of the city, enabled architects and city planners to redesign the city, making it more organized and fire resistant. David Cowan is an author, a veteran firefighter, and an award-winning journalist who has written for major newspapers and magazines and appeared in numerous television documentaries about historic wildfires. He lives in Chicago.*

Nineteenth-century America saw almost every major city come close to burning down, including New York, Philadelphia, Pittsburgh, Boston, and Portland, Maine. None stirred the nation like the Chicago Fire, which still measures the city's unique will and identity. Chicago may have been incorporated in 1837, but events in its timeline are qualified as having occurred either "before" or "after" the 1871 fire, making it the single most significant event in the city's history. The fire gave city planners and architects a clean slate on which to redesign a new city plan far more coherent than the original one that had sprouted up erratically and hastily in a time of largely unregulated boom. It also made possible the gem that Chicago is today with its open lakefront and systematic grid of streets, boulevards, and parks. . . .

David Cowan, *Great Chicago Fires: Historic Blazes That Shaped the City.* Chicago: Lake Claremont Press, 2001. Copyright © 2001 by David Cowan. All rights reserved. Reproduced by permission.

The fire's universal appeal [is] the basis of one of America's most famous folk tales: that of Mrs. O'Leary's disgruntled cow kicking over the lantern. That the fire began in the O'Leary barn at Jefferson and DeKoven Streets is beyond dispute. But the rest of the story is now generally accepted as myth, fabricated by an unscrupulous newspaper reporter who capitalized on anti-Irish sentiment. Even though the Chicago City Council passed a resolution in 1998 formally absolving them of any responsibility, the O'Leary clan and their tragic matriarch Catherine were never quite able to escape the fire's stigma. Historians now believe it was probably an O'Leary neighbor who sparked the most famous fire in U.S. history.

A Fire-Prone City

By 1871, Chicago had grown into a major American city, the Queen of the West, thanks to the I&M [Illinois and Michigan] Canal and the railroads. In just 40 years it had gone from quiet prairie outpost to a metropolis exceeding 334,000, the U.S.'s fourth largest. But if any city in America was destined to burn, Chicago was it. Between 1858 and October 1871, the city experienced nearly 3,700 fires, eight of which were major loss blazes. . . . This alarmingly high number of fires, 700 occurring in one year alone, prompted Lloyd's of London to send a representative who issued a report that convinced the firm to cease insuring property in Chicago and to cancel its outstanding policies.

The drastic action was justified. Unlike London, which had evolved over centuries (and burned in 1666), Chicago was thrown up in a hurry. By 1871 it had expanded to 23,000 acres, nearly 36 square miles, with city limits extending west to what is now Pulaski Road, north to Fullerton Avenue, and south to 39th Street, containing 60,000 buildings, 90 percent of them wood. This concentration of closely-built wooden structures, including hundreds of miles of wooden streets, sidewalks, and picket fences, made Chicago a huge tinderbox. As the center of the country's woodworking industry its enormous fire potential was increased by an abundance of furniture mills, lumber yards, and paint and varnish shops, each combining to churn out thousands of wood-based products. Chicago also boasted the largest grain market in the world, storing its vast supply in 17 massive grain elevators built along the Chicago River, a waterway with 24 bridges. Several of the city's lumberyards dotted the riverbanks, whose dock areas were often clogged with up to 200 wooden sailing vessels. Despite this impressive growth, the city's fire department had failed to keep pace. The department employed only 216 men, including fire alarm operators, and maintained a small fleet of just 17 steamers and four hook-and-ladders, less than one engine per square mile.

The summer and fall of 1871 were hot and dry in the Midwest, causing a severe drought that left much of the region open to numerous grass and forest fires. Rainfall had only been one quarter the normal amount: only one inch had fallen since July, making the wooden city of Chicago ripe for disaster.

The Prelude of the Great Fire

On Sunday, October 8, Chicago's understaffed fire department was exhausted. Since September 30, it had battled 20 major fires, four of them multiple-alarm blazes. The worst of these had occurred the night before. This so-called "Saturday Night Fire" burned more than four square blocks in the West Division, destroying property valued at $750,000. Though Chicago's largest fire to date, it would be eclipsed by the happenings of the following day.

The Saturday fire began between 10 and 11 P.M. at the large Lull and Holmes planing mill at 209 South Canal Street in the boiler room. The mill stood in the center of the block next to several lumber sheds, coal yards, wooden outbuildings, and smaller houses and shanties. By the time the first fire steamers and hose carts pulled up, the mill was almost entirely engulfed in flames. A strong south wind quickly spread the fire to neighboring buildings, and in less than 20 minutes all of Canal Street between Jackson and Harrison was ablaze, as was Jefferson Street. The fire traveled east toward the South Branch of the Chicago River, devouring everything in its path. On the north, firefighters mounted a heroic fight that prevented it from extending beyond Adams Street. But the 17-hour battle had taken an enormous toll on both men and equipment. Along with lesser equipment, the fire had destroyed Pioneer Hook-and-Ladder Co. 1 and numerous lengths of hose, and severely depleted the department's supply of coal needed to power its pumping steamers.

The next day there were about 125 firefighters on duty, many of them badly in need of rest. Especially fatigued were several fire companies who had worked up to 22 hours straight. But if the men were hoping for a brief respite, they were in for a big letdown. Though they didn't know it yet, the previous night's fire was merely a prelude to the approaching apocalypse. At around 8:40 P.M., flames broke out about four blocks south of the Saturday night fire in the cattle barn behind 137 (now 558) West DeKoven Street owned by Patrick and Catherine O'Leary. The O'Learys were a working-class family with five children; Patrick was a laborer, while his wife ran a neighborhood milk route. Together they owned five cows, a calf, a horse and a wagon, and property consisting of two frame houses and a barn in the back. The O'Learys lived in the smaller house and rented the larger to another family, the McLaughlins. The O'Leary barn backed up to a common alley and the fire spread rapidly beyond it, catching on to other adjacent barns and shacks. Six blocks away inside the Maxwell Street firehouse of Engine 6, *The Little Giant*, weary firefighters were trying to get some well-needed rest when Joseph Lauf, the watchman on duty, spotted the fire from his perch in the tower. "Turn out," Lauf yelled to the other six men in his company. Led by their foreman, Bill Musham, the crew quickly hitched their team of five horses to the company's Amoskeag steamer and hose cart. After starting the steamer boiler, they pushed out for the fire at DeKoven and Jefferson.

At 9:16 P.M., the spotter inside the courthouse tower north of the river saw the fire as well and struck Box 342, giving an inaccurate location of Halsted and Canalport, more than a mile southwest of the fire's position. Just as the signal was ringing across the fire department telegraph system, a passerby ran into the Jefferson Street firehouse of Engine 5, *The Chicago*, and Truck 2, *The Protector*, announcing that a fire was burning a few blocks south, opposite the location given by the fire alarm office. But it no longer mattered because by now the fire had grown so big that responding companies could see the glow in the sky and went straight for it. (Some minutes before, an O'Leary neighbor ran over to a nearby drugstore to turn in an alarm, but when the proprietor inserted his key and pulled the alarm box, the signal was never received at fire alarm headquarters. Three other alarm boxes in the neighborhood were not pulled until much later.)

Chicago's Fate Was Sealed

Unfortunately, lost time and a fierce south wind had already sealed the fate of Chicago. Engine 6 arrived at 8:45 P.M. and got "first water" on the fire south of DeKoven Street that by now involved up to 30 buildings and shot flames 60 feet into the air. The fire, however, driven by hot blowing winds, was pushing north where Engine 5 set up and tried stopping it. But when its steamer ran out of coal, the flames jumped Taylor Street and spread from wooden building to wooden building. When Engine 5's pump ground to a halt from a lack of steam pressure, frustrated firemen tried stoking its boiler by ripping up boards from the sidewalks and fences, and even ran back several blocks to their engine house at Van Buren and Jefferson to get buckets of coal. But the fire didn't wait. It swarmed north across Taylor Street toward St. Paul's Church, feeding on the roofs and walls of dozens of buildings before turning into one huge sweeping arm of flame that was now poised to consume everything in its path.

The fire in the West Division did not extend west of Jefferson Street or south of 12th Street. But the South Division, which included the central business, financial, and mercantile districts as well as Chicago's premier hotels, museums, and office buildings, was about to be reduced to ash. The same fate awaited the residential North Division containing the city's finest homes and mansions. Gale-forced winds turned the mounting blaze into a firestorm, and as the clock ticked away the rest of that Sunday night and Monday morning, the fire line expanded rapidly and the "Lightening City" of Chicago steadily disappeared. Throughout the burning districts, flying embers known as "fire devils" danced from rooftop to rooftop, turning night into day. Spires on magnificent churches ignited, then crumbled and crashed to the ground. Factories burst into flames. Sheds, barns, homes, and shanties were obliterated. Covering the entire city was a towering wave of orange that smelled of burnt wood and flesh. Said one fireman: "you couldn't see anything over you but fire . . . that night the wind and the fire were the same."

Chicago had become another Rome, though no one fiddled as the city burned. No longer could the fire be stopped, only watched. By 1:30 Monday morning, it had reached the city's gas works, though thanks to a brave engineer most of the gas had been diverted into sewers and a reservoir. By 2 A.M. the courthouse, where President [Abraham] Lincoln's body had lain in state six years earlier, began to burn. More than 100 prisoners in its basement were freed and told to run for their lives. Five others convicted of murder were handcuffed and taken away by police. With its windows melting and masonry crumbling, the roof fell in, including the huge bell inside the tower, which on its way down destroyed the fire alarm office on the third floor.

In their panic to escape the city, people trampled one another. Even as they tried to outrun the flames, women stopped to give birth in the street, their labors induced by fear and excitement; some mothers and newborns burned to death on the spot. Many residents jumped in Lake Michigan or ran to the outlying prairie. Others sought refuge in open graves of the city's cemetery, which at the time was being moved from Lincoln Park. When flames easily jumped the main branch of the Chicago River, it put to rest all claims that the river would act as a natural fire barrier. The army's attempts to blow up buildings in the fire's path also proved useless as fire stops. By 3:30 A.M., the gothic waterworks at Michigan and Chicago Avenues, five miles north of the O'Leary home, caught fire, knocking out all of the city's remaining pumping capability. When the water mains ran dry, hope was lost for saving the North Division.

The Aftermath of the Fire

By 7 A.M. Monday, factories like McCormick's reaper works lay in a heap of ruins. All bridges over the river's south branch were destroyed while only two remained to take people into the North Division. Train depots and railroad cars were razed, but not before fire engines arrived by railway from other Illinois cities, as well as Milwaukee and Cincinnati. This equipment was sent into the North Division, where fire seared the homes of millionaires. In the end, more than 13,000 homes were consumed in the North Division, where the fire had advanced to Lincoln Park and as far north as Fullerton Avenue. But it wasn't out-of-town fire equipment that finally brought the great fire to a halt. With nothing left to burn, upon reaching the prairie, it simply burned out. The last house destroyed was owned by Dr. John H. Foster along Lincoln Park. One more house north of the city limits did catch fire but was not considered within the burnt district. At 3 A.M. on October 10, a strong downpour extinguished whatever hotspots remained.

Chicago looked like a scorched wasteland. The fire had consumed a three-and-a-half square mile area—from Harrison Street north to Fullerton Avenue, and from the river east to Lake Michigan. Lost were some 15,700 buildings, including the entire central business section. About 300 people died in the fire, and 100,000 were left homeless. The exact number of deaths remains uncertain because many victims were burned beyond recognition. Others drowned in the river. Damage was estimated at $200 million, of which only $88 million was recovered. (After the fire, 60 of Chicago's estimated 250 insurance companies went bankrupt.) A few priceless items were also lost, including the orignal draft of President Lincoln's Emancipation Proclamation when the Chicago Historical Society burned. Also gone were valuable city, county, and business documents and records. Somehow, the wooden Ogden mansion on what is now the site of the Newberry Library, survived. So did the frame house of Patrick and Catherine O'Leary, but not their barn.

The San Francisco Fire of 1906

By Gerstle Mack

Gerstle Mack, an architect and author, was born in San Francisco in 1894 and was not quite twelve years old when an earthquake and fire struck the city on April 18, 1906. By the time Mack wrote 1906: Surviving San Francisco's Great Earthquake and Fire *in 1981, he was one of the few survivors who were old enough during the disaster to remember it vividly. In this excerpt from his book, Mack tells how a number of small fires started burning at once just minutes after the earthquake. The quake had cracked chimneys, allowing sparks to ignite roofs and toppled electric wires onto houses. The quake had also broken gas mains, which subsequently exploded into flames. In less than an hour after the earthquake, more than fifty buildings were burning. Unfortunately, firefighters were unable to extinguish the fires because the quake had destroyed the city's main water lines. The author describes how during the first twenty-four hours this fast and unstoppable fire consumed the city's financial and market districts, mansions, and hotels. Firefighters attempted to stop the conflagration using dynamite to blow up buildings in hopes of reducing the fire's fuel, and by spraying the fire with salt water pumped from the bay through mile-long hoses. In the end, the earthquake and fire destroyed about twenty-eight thousand buildings and killed approximately three thousand people.*

Within fifteen minutes after the earthquake, ominous columns of smoke could be seen rising in different parts of the city. The fires started in various ways: stoves were lighted in houses whose owners were unaware that their chimneys had cracked, and sparks set the roofs ablaze; gas exploded in mains broken by the earthquake; electric connections were damaged, causing short circuits; high-tension wires snapped and fell across shingled roofs. Eyewitness accounts vary so greatly that the precise number of "original" fires is uncertain, but there were at least a dozen, perhaps twenty. Less than an hour later more than fifty buildings were aflame. Most of these were in the district south of Market Street, where several separate fires soon merged into a conflagration with its center in the vicinity of Third and Mission streets. The fire department was efficient and well equipped, but the firemen were handicapped at first by the scattered locations of the burning buildings, which obliged them to dissipate their efforts over a wide area. In a very short time another and infinitely more serious difficulty became apparent: there was no water. The streams from the hoses diminished to mere trickles, then stopped altogether. A few roof tanks, underground cisterns, artesian wells, and small local reservoirs continued to supply water, and in some flat sections of the city water was pumped from sewers; but the earthquake had broken most of the mains within the city as well as the big conduits leading from the great reservoirs outside. The firemen were almost helpless.

At that time, before the construction of the present Hetch Hetchy reservoir in the Sierras, San Francisco derived its water supply from four sources: the San Andreas, Crystal Springs, and Pilarcitos lakes south of the city, and one or two reservoirs in Alameda Country across the bay. The conduit running under the bay from the east was undamaged, but breaks in the pipes on shore at Dumbarton Point put this system out of commission. The Crystal Springs conduit, partly laid in very soft ground, was wrenched and fractured in many places, the worst destruction occurring where the line crossed a salt marsh on a trestle between San Bruno and South San Francisco. From Pilarcitos the path of the conduit coincided for about six miles with the actual line of the great fault; throughout this section the pipes were torn apart violently at a number of points, as violently telescoped at others. The San Andreas pipeline, passing through fairly solid ground approximately midway between the fault and the marsh, suffered relatively slight damage and was repaired within three days after the earthquake—too late to save the city from the flames but in time to avert a serious water famine. For several weeks this reservoir supplied all the water that flowed into San Francisco. Within the city limits smaller storage reservoirs held about eighty million gallons, but although these reservoirs remained intact the water could not be distributed because many of the mains and branch pipes, especially those laid in soft earth, were broken, and the system lacked bypasses through which water might have been detoured.

Fires Begin and Spread Quickly

Early in the morning a series of scattered fires near the waterfront quickly combined into another center of conflagration. East Street (now called the Embarcadero), a very broad curving thoroughfare separating the city from the docks that jutted into the bay, was flanked on its west or city side by rows of old wooden buildings housing the tawdry establishments common to all seaports: ship chandleries, drab restaurants and saloons, frowzy hotels and boarding houses catering to sailors, "cheap John" secondhand clothing shops. The flames consumed these rattletraps like tinder and swept on through the adjacent factory district, more substantially constructed but not much more resistant to the intense heat of the blaze. At first there seemed to be a faint chance that the great width of Market Street, the main thoroughfare of the city, might suffice to confine the conflagration to the southern areas, but this hope quickly faded. The ramshackle wooden buildings on East Street north of Market, similar to those a little farther south, soon caught fire. From these the flames spread westward through the wholesale district and north towards the Latin Quarter at the foot of Telegraph Hill.

About ten in the morning a new fire started independently near the corner of Gough and Hayes streets, in a middle-class residential section known as Hayes Valley not far from the present Civic Center, and advanced eastward across Van Ness Avenue to sweep through the already ruined City Hall. This blaze was nicknamed the "ham-and-egg" fire because it was said to have been started by a woman who, not realizing that her chimney was cracked, cooked herself a breakfast of ham and eggs and set fire to her kitchen wall and her roof. Meanwhile the flames drove swiftly along Mission Street, roared through the Grand Opera House, and attacked the great office buildings, stores, and hotels on the south side of Market Street. . . .

Attempting to Stop the Fire

There had been a number of attempts to halt the conflagration by dynamiting buildings in various parts of the city, but the firemen, inexperienced in the use of explosives, hesitated to demolish areas sufficiently broad and continuous to check the flames,

and their timid efforts were ineffectual. Moreover, the supply of
dynamite on hand was limited. Early on Wednesday morning the
acting fire chief sent a request to the military post at the Presidio
for all available explosives and for a detail of troops to handle
them. Three hundred pounds of dynamite and several barrels of
artillery powder were sent at once; a little later, larger amounts of
dynamite were obtained from the California Powder works
(which afterwards refused to accept any payment for this contribution); and a small quantity of gun-cotton arrived from the
naval base at Mare Island. On Thursday afternoon a westerly
breeze again started to blow, this time briskly enough to reduce
the speed of the fire's advance to about half a block an hour. As
the flames crept slowly down the western slope of Nob Hill the
fire fighters decided to make one last desperate attempt to save
the Western Addition, the principal residence district of the city,
by boldly dynamiting and backfiring all the buildings—mostly
dignified old-fashioned wooden dwellings—on the east side of
Van Ness Avenue, in the hope that the space thus cleared, plus
the great width of the avenue itself, would stop the westward
progress of the fire. The charges were set by the firemen, de-
tachments of soldiers from the Presidio, and volunteers; fuses
were lighted, and row after row of fine houses blew up with a
deafening roar. Other houses along the avenue were sprayed with
kerosene and backfired, and field guns battered down all walls
and fragments of walls left standing after the blasting and burning had done their work.

This belated campaign of deliberate ruthless demolition was,
to a very great extent, successful. The fire was checked on the
eastern side of Van Ness Avenue except in the five blocks between Clay and Sutter streets, where it crossed the broad thoroughfare, consumed all of the houses on the opposite side (including the magnificent red sandstone residence of Claus
Spreckels), and swept one block farther west to Franklin Street.
Here, with the aid of salt water pumped from the bay through
hoses a mile long, and with the additional assistance of the west
wind, the exhausted fire fighters finally succeeded in extin-
guishing the flames at about two o'clock on Friday morning,
forty-five hours after the start of the conflagration. At approximately the same hour, the fire in the Mission district was halted
near 20th Street on the east side of Dolores, a street as wide as
Van Ness Avenue and an even more effective barrier to the

flames. The Western Addition and the more remote, more thinly populated residence areas beyond, extending westward to the ocean, were safe, as were all of the poorer sections of the city south of 20th Street, and the Potrero district with its iron foundries and other heavy industries.

The Final Battle

But the fire, stopped on three sides, still blazed fiercely on the fourth. The flames sweeping over Nob Hill had remained south of Washington Street until Thursday afternoon, when they jumped across the narrow street and spread rapidly to the north. Fanned by the west wind which had helped to slow down and finally to extinguish the main conflagration, the new blaze advanced in the opposite direction—eastward—and climbed the western slope of Russian Hill, then over the crown and down the other side into the thickly settled valley of the Latin Quarter, and up Telegraph Hill. The inhabitants of the North Beach district, caught between the fire and the bay, were rescued by tugs and steam schooners. A few houses on top of Russian Hill were saved by their owners, assisted by neighbors, who filled bathtubs and other receptacles with water drawn from a small local reservoir and, for seven hot and weary hours, stubbornly beat out the flames with wet blankets. Their strenuous efforts were considerably hampered by what the *Chronicle* called "the almost irrepressible desire of the dynamiters to blow up the entire row." On Telegraph Hill several houses were preserved by similar methods, but here there was little or no water, so the inhabitants, most of whom were of Italian birth or descent, soaked their blankets in casks of wine.

All day Friday and on Friday night the fire raged over Russian and Telegraph hills and through North Beach. By Saturday morning it had reached the bay, had swept through the lumber yards and iron foundries located at the foot of the hills, and was threatening Pier 27, the northernmost of all the docks. Here the worn-out fire fighters made a final stand. Dousing the pier with salt water from the bay, they quenched the last blaze in this area. By mid-afternoon on Saturday, April 21—three and a half days after the earthquake—the fire was out. For more than three weeks, coal dumps in the industrial district south of Market Street continued to burn, and piles of coffee and tea in a warehouse smoldered, emitting a rich aroma; but these half-smothered

embers were no longer sources of danger, because everything else
in their victinity had already been destroyed.

The Fire's Legacy

The fire was out; but most of the city was in ruins. The area of
the burned district covered 2,593 acres—a little more than four
square miles—comprising 490 city blocks and parts of 32 others. The flames had devoured 28,188 buildings, and the irregular
perimeter of the ruined section measured more than eleven miles
in length. The area of destruction was eighteen times that caused
by the Baltimore fire of 1904, eight times as great as that of the
famous fire of London in 1666, and one-fourth larger than the
area of the Chicago fire of 1871. In San Francisco the property
loss was estimated at about five hundred million dollars, or $1,100
per capita. In terms of dollars and cents it was by far the greatest
disaster that had ever befallen a single city. All of the wholesale,
retail, and financial areas were destroyed. Every good hotel was
gone, and most of the cheaper ones had vanished as well. Not a
theater or restaurant was left. Practically every store larger than a
neighborhood grocery was in ashes. The municipality of San
Francisco, which was conterminous with San Francisco County,
sustained very heavy losses. The City Hall, the Hall of Records,
the Hall of Justice, the county jail, the public library (housed in
a wing of the City Hall), thirteen fire engine houses, and numerous police stations were burned. Twenty-nine public schools
were totally wrecked by the fire, and many others constructed—
as the City Hall had been—by incompetent or dishonest con-
tractors, were severely damaged by the earthquake.

Added to the financial loss was the destruction of irreplace-
able records. All certificates of births, marriages, and deaths were
burned. Of more than fourteen hundred thick volumes of mort-
gage registrations in the Hall of Records, only a single volume
was salvaged. Three-fifths of the property deeds went up in
smoke. On the other hand, quick-witted and courageous employees carried away and preserved many valuable documents:
the tax rolls, the papers and photographs of the police department, and the ancient records, priceless to historians, of the days
of Spanish and Mexican sovereignty. Damage to powerhouses
and transmission lines had put out of commission all electric
lights, telephones, telegraph wires, and streetcars (at that time
mostly cable-cars) in the city. There was no gas; the earthquake

had broken the mains. The city had little water and less food. About two hundred thousand people—almost half the total population—were homeless. . . .

Human Losses

Considering the magnitude of the disaster, the number of people killed or injured was remarkably small. The figures varied slightly in different accounts but all agreed that the city's death list totaled about five hundred. Probably the most reliable estimate was that given by Maj. Gen. Adolphus Washington Greely, commander of the Division of the Pacific, in his official report dated July 30, 1906. According to this computation, 498 people were killed (including those who died later of injuries received at the time of the catastrophe) and 415 were seriously hurt but recovered. Among the dead, 194 bodies, most of them charred beyond recognition, were never identified. On the first day many corpses were hastily buried in shallow trenches dug in Portsmouth Square before any serious attempt at identification could be made; and by the time they were disinterred, several days later, and transferred to cemeteries for permanent burial, it was too late. Nearly all of the deaths were caused, directly or indirectly, by the first earthquake shock; the victims were either killed instantly by falling walls and chimneys or pinned so firmly under heaps of debris that they could not be extricated before the fire reached them. Hundreds of others, injured and half buried in rubble, were saved by heroic rescuers who, often at the risk of their own lives, dug and hacked their way through the ruins and dragged out the helpless survivors.

One reason for the relatively short death roll was the resilience of the wood-framed houses in which most San Franciscans lived. Another was the early hour; at five in the morning few people were outdoors, and as much more debris toppled into the streets than into the dwellings, the inhabitants were safer at home than anywhere else. San Francisco's 500 fatalities, in addition to approximately 250 in other communities in the earthquake zone, amounted to a small fraction of the toll which has been taken by other great earthquakes: more than 80,000 killed at Messina [Italy] in 1908, for example, and at least 145,000 at Tokyo [Japan], where fire accounted for more deaths than the earthquake, in 1923.

China's Great
Black Dragon Fire

By Harrison E. Salisbury

*On May 6, 1987, a forest worker in northern Manchuria spilled
some gasoline while he was refilling his handheld brush cutter. A spark
from the cutter ignited the spill and started a wildfire so large that the
smoke could have been seen from the Moon. The Great Black Dragon
Fire burned vast forested areas in China and Russia for almost a month
and destroyed valuable timber covering an area roughly equal to the size
of New England.*

*In spite of its monumental dimensions and impact on the Chinese
timber industry, this disaster went largely unnoticed in Western countries.
One of the reasons was that no foreign correspondents were allowed to
visit the area during the fire. Only Harrison E. Salisbury, whose long-
standing specialization in Russian and Chinese issues earned him
China's permission to visit the region after the fire, was able to bring to
the West accounts of the disaster. In this selection from his book* Great
Black Dragon Fire: A Chinese Inferno, *Salisbury describes the burning of the city of Xilinji. He describes how Political Director Wang Aiwu
tried to save the lives of thousands of villagers by allowing them to take
refuge in his military compound on the edge of the city. Unfortunately,
the fire began to torch buildings within the compound, threatening to ignite stores of ammunition and fuel. Before such a catastrophe could occur,
Wang evacuated the refugees, all of whom survived. Harrison E. Salisbury is a journalist and author of more than twenty books; he has received numerous awards, including the Pulitzer Prize.*

In a state of chaos, each man or woman sees only a fragment
of the whole. So it was for the Black Dragon kingdom in the
time of the fire. The people in one part of Xilinji did not
know what was happening in another. The people in Tuqing, the

next town to the east, did not know what was going on in Xilinji. . . .

In the Fire's Path

As the wind moved from west to east, the fire snapped commu-nications as if they were beads on a chain, cutting telephone and telegraph lines. . . . The whole forest apparatus effectively [was] blinded.

The first point to be lost was Xilinji, the western anchor. It went out at 7:10 P.M. Next was Tuqing, forty miles east of Xil-inji. It was lost at 8:00 P.M. Amur, fifteen miles farther east, hung on until 10:10 P.M. . . .

The town of Xilinji is shaped like a slightly elongated box. Be-cause of Xilinji's location along the Tang Ta River, a tributary of the Emur, its streets run not north-south and east-west but northwest-southeast and northeast-southwest. The Tang Ta River lies north and east on the outskirts of the city, and a bridge con-nects the city with the gravel road to the Black Dragon River and Mohe, about fifty miles to the northeast. The east-west rail-road spur runs along the western edge of the city and then turns southwest to the Gulian forest farm.

To the south of the city there is a low range of hills from which a panoramic view of the town can be had. There are also hills along the northwestern fringe of town, and it was from these hills that the fire descended on Xilinji. The principal buildings— the railroad station, post office, School Number 1 (a high school), government and party offices, guesthouse, hospital, and banks— were located in the northern half of the city. The school was in the northwest at a crossroads facing a wide, open square where the free market in meat, vegetables, clothing, and secondhand goods had been located since Deng Xiaoping's relaxation of pri-vate trade and commerce. Midway along the western flank of the city stood the railroad station and post office. Close to the rail-road before the fire stood the principal timber depot with tens of thousands of logs and stored timber awaiting shipment. On the southeastern outskirts of the town, not far from the Tang Ta River, sprawled the Ninth Regiment military compound.

The rest of town was given over to small shops and housing, some private, some barracks, almost all one-story wooden struc-tures. Near the center of town was a pleasant park of virgin conifers, uncut and left as a monument to the forest that once grew on the spot where the town was founded in 1974.

When I saw Xilinji this geography was difficult to reconstruct. The entire city had been destroyed by the fire except for half a dozen brick and concrete buildings of three or four stories. These structures were surrounded by devastation. Intensive construction was going forward. Thousands of carpenters labored to get housing up before the onset of early winter.

The Fire Reaches Xilinji

On May 6 Political Director Wang Aiwu was in charge of the Ninth Regiment of the PLA [People's Liberation Army] because his two seniors, the commander and the deputy commander, were in Harbin undergoing medical treatment. Most of the troops were on duty along the border.

On the morning of May 7 Director Wang had been briefed on [the fire at] Gulian. The fire was under control, but a shortage of food for the fire fighters had developed. The director drew supplies from the regimental stocks and had them on their way in two hours.

The Ninth Regiment compound was a spacious area in southeastern Xilinji, comprising a large courtyard bigger than a parade ground, commodious barracks, an excellent headquarters building, four other good-sized buildings, and big ammunition and oil depots. . . .

At 6:00 P.M. on May 7 Director Wang saw from his courtyard that the wind was rising fast and the town was becoming shrouded in smoke. He estimated the wind at just short of 50 miles an hour.

"I witnessed this," the director said, a little dramatically. "Within thirty or forty minutes the situation had become terrifying. The fire was beyond control. The need was to save the people. The fire might be compared to a typhoon, the waves were coming at express speed. The situation was one of terror and danger. It was unlike anything I had seen."

Amid all this he was, he said, tremendously impressed with the calm and heroism of the county leaders and forest men. They kept working quietly on a firebreak with the flames only a block distant. But they could not stop the fire.

He took a detachment of a hundred men into town and led them to the western highway. The streets were almost impassable, and he decided to go back to protect his headquarters and the ammo and oil depots, to mobilize his men, and to turn the compound into a refugee center. By the direction of the wind Wang knew the compound would be in the fire's path. He hoped its great expanse of open ground would save it from destruction. His men had to fight their way through the crowds streaming from the city. Back at the compound he found his men scattered throughout the area, and he had trouble rounding them up. Ordinarily he summoned them with a bugle call. But this was now impossible—the call was recorded on tape, and the electricity was out.

He had one concern. If his ammunition and petrol dumps blew up, the blast would equal a small atom bomb, enough to wipe out headquarters and the city.

Wang decided to take a chance. He ordered the gates opened. People swarmed in. "They didn't know what they were doing," he said. One woman was carrying her baby upside down. The child's face was blanched. He put the child right side up. Women clutched at his hands. He told them to stop running, find a place on the ground, and sit down. A lady in her seventies cried: "What do I do now?" He said: "Calm down. Go sit on that bench and keep your mouth open." An hour later he came by. She was still sitting there, her mouth open. He told her she could close it now.

Director Wang had taken a calculated gamble when he opened the compound gates. The fire had come down from the northwestern hills and was cutting a broad swath through the town from northwest to southeast. The compound lay on the far eastern outskirts of town, but as the wind brought the flames into town they touched off the wooden buildings, broadening the path of fire and moving it closer and closer to the big compound.

Wang still had a line open to the military command of the Black Dragon area. He took the precaution of asking for instructions. The answer came back: give priority to protecting the ammunition and oil depots. Defend them against destruction at all costs.

Protecting the Ninth Regiment Headquarters

Soon the fire began to threaten both refugees and depots. A fireball from the northwestern hills rocketed into the third floor of the headquarters building, and the floor exploded in flames. Sparks and fiery debris catapulted into the camphor trees that shaded the compound.

Within half an hour of Wang's decision to let the people into the compound, it became clear that they could be trapped in a whirlwind of flame. The ammunition dump was located fifteen hundred feet away at the southwestern edge of the big compound. Embers lodged in a pile of creosoted logs beside the ammunition dump, a brick and concrete vault fitted with an iron door and packed with tons of shells, mines, grenades, fuses, and bullets. The fire fighters threw sand on the logs. It didn't help much.

Somehow wind whipped open the iron door beside the blazing logs. Three soldiers ran to close it. Two were driven back by the pitch-laden smoke. The third poured water over his cotton padded jacket and made his way to the clanging door. He managed to slam it closed at the cost of burned hands and face. He was one of thirty-two men in the regiment to be decorated for bravery.

Fire moved closer and closer to the five thousand refugees assembled on the parade grounds. When it began to leap from one camphor tree to another, Director Wang sent in a team of soldiers with axes. They cut the trees, one by one, those burning, those yet unburned. When the brick building that housed headquarters began to burn, Wang ordered soldiers to wet blankets and hang them over the windows of the upper stories to prevent sparks and embers from spreading the flames from floor to floor. The wind whipped the blankets from the hands of the men. Some soldiers were almost hurled to the ground.

Hardly had the danger to the ammo depot been averted when the asphalt roof of the oil-storage vault, at the southeast corner of the compound, caught fire. Troops dampened their cotton comforters and used them to douse the blaze. Then they draped more wet comforters over the asphalt to catch the rain of sparks carried by the wind.

Other soldiers jumped into water reservoirs, soaking their jackets, then rolled over and over in the flaming grass to keep fire out of the parade quadrangle, where the refugees were clustered. Later the men got a "Rolling Over the Fire" medal, created specially for the occasion.

As fire surrounded the parade grounds, sparks flying overhead, flames shooting from buildings, smoke enshrouding the scene, pandemonium broke out. Women had brought all of their possessions they could carry on their backs—TV sets, radios, bedding, pots and pans, cotton sacks of valuables. Now they feared they would lose everything, including their lives. The women shouted. Babies howled.

Wang consulted the city authorities. He had real fear that the flames—hardly a hundred yards away—might break through. The ammunition might blow up or the gasoline catch fire.

It was decided to truck the people to safer spots. It was now close to 9:00 P.M. The fire had passed over the river. Many refugees were taken to the water's edge. Others were carried by a roundabout route to the highway crossroads beside School Number 1, which had become the main assembly point for burned-out families.

When all the refugees had been removed, Wang began to assess the results. He and his men had saved the lives of more than five thousand people. Not one of those he had taken in had been injured. Nor had any of his men lost their lives. True, while they were caring for the refugees, their own barracks and houses in the compound had burned. They had lost all they had: furniture, color TV sets, electric refrigerators, bedding, clothing. But no one lost a wife or child. No one in the command had been seriously injured. The worst damage was scorched skins and hands, including those of Director Wang.

The Great Yellowstone Fire

BY THE U.S. NATIONAL PARK SERVICE

The fires that burned Yellowstone National Park in Wyoming in 1988 were the most severe to strike the area since the park's creation in 1872. The fires consumed 36 percent of the park, damaged an estimated $3 million worth of property, and killed numerous animals. In this selection the National Park Service (NPS) explains that fire has always played a role in the ecology of Yellowstone National Park. Fire is essential for the reproduction of some plants in the park, like lodgepole pines, which have cones that only release their seeds after the intense heat of a fire melts the resins that keep the cones sealed. The service points out that although fire is destructive, it is also necessary for the periodic renewal of land and wildlife. Evidence for this, the NPS contends, is the condition of Yellowstone just a few years after the blaze. Striking displays of wildflowers have covered burned areas and the whole forest is engaged in a new cycle of growth. This selection is published on the official website of Yellowstone National Park.

L andscapes such as those seen in Yellowstone have long been shaped by fire and not just the cool, creeping ground fires often described as "good" for grass production. The natural history of fire in the park includes large-scale conflagrations sweeping across the park's vast volcanic plateaus, hot, wind-driven fires torching up the trunks to the crowns of the pine and fir trees at several hundred-year intervals.

Fire History

Such wildfires occurred across much of the ecosystem in the 1700s. But that, of course, was prior to the arrival of European explorers, to the designation of the park, and the pattern estab-

U.S. National Park Service, "Wildland Fire," www.nps.org, October 2002.

lished by its early caretakers to battle all blazes in the belief that fire suppression was good stewardship. Throughout much of the 20th century, park managers and visitors alike have continued to view fire as a destructive force, one to be mastered, or at least tempered to a tamer, more controlled entity. By the 1940s, ecologists recognized that fire was a primary agent of change in many ecosystems, including the arid mountainous western United States. In the 1950s and 1960s, national parks and forests began to experiment with controlled burns, and by the 1970s Yellowstone and other parks had instituted a natural fire management plan to allow the process of lightning-caused fire to continue influencing wildland succession.

Many of Yellowstone's plant species are fire-adapted. Some (not all) of the lodgepole pines (*Pinus contorta*), which make up nearly 80% of the park's extensive forests, have cones that are serotinous sealed by resin until the intense heat of fire cracks the bonds and releases the seeds inside. Fires may stimulate regeneration of sagebrush, aspen, and willows, but the interactions between these plants and fire is complicated by other influences such as grazing levels and climate. Though above-ground parts of grasses and forbs are consumed by flames, the below-ground root systems typically remain unharmed, and for a few years after fire these plants commonly increase in productivity.

Letting Some Fires Burn

In the first sixteen years of Yellowstone's natural fire policy (1972–1987), 235 fires were allowed to burn 33,759 acres. Only 15 of those fires were larger than 100 acres, and all of the fires were extinguished naturally. Public response to the fires was good, and the program was considered a success. The summers of 1982–1987 were wetter than average, which may have contributed to the relatively low fire activity in those years.

No one anticipated that 1988 would be radically different. In April and May, Yellowstone received higher-than-normal rainfall. But by June, the greater Yellowstone area was experiencing a severe drought. Forest fuels grew progressively drier, and the early summer thunderstorms produced lightning without rain. The fire season began, but still without hint of the record season to come. Eleven of 20 early-season fires went out by themselves, and the rest were being monitored in accordance with the existing fire management plan.

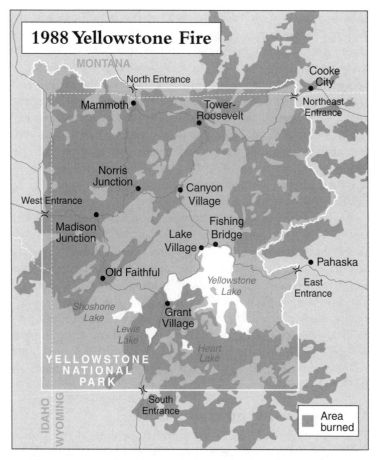

The summer of 1988 turned out to be the driest in the park's recorded history. By July 15, only 8,500 acres had burned in the entire greater Yellowstone area. Still, due to continued dry conditions, on July 21 by which time fire activity had become noticeable to park visitors and to the national media the decision was made to suppress all fires. But within a week, fires within the park alone encompassed more than nearly 99,000 acres, and by the end of the month, dry fuels and high winds combined to make the larger fires nearly uncontrollable. National news reporters poured into Yellowstone National Park, as did firefighters from around the country, bolstered by military recruits. On the worst single day, August 20, 1988, tremendous winds pushed fire across more than 150,000 acres. Throughout August and early September, some park roads and facilities were closed to the public, and residents of nearby towns outside the park feared for their property and their lives. Yellowstone's fire management policy was the topic of heated debate, from the restaurants of park border towns to the halls of Congress.

After the Fire

By September 11, 1988, the first snows of autumn had dampened the fires as the nation's largest fire-fighting effort could not. The imminent danger to life and property was over, and firefighters were gradually sent home, although the last of the smoldering flames were not extinguished until November. Staff in Yellowstone National Park went to work surveying the impacts of the fires on wildlife, plants, historic structures, trails, and more, and answering the demands for information, explanation, and a new fire management policy.

A total of 248 fires started in greater Yellowstone in 1988; 50 of those were in Yellowstone National Park. Despite widespread misconceptions that all fires were initially allowed to burn, only 31 of the total were; 28 of these began inside the park. In the end, 7 major fires were responsible for more than 95% of the burned acreage. Five of those fires were ignited outside the park, and 3 of them were human-caused fires that firefighters attempted to control from the beginning. More than 25,000 firefighters, as many as 9,000 at one time, attacked Yellowstone fires in 1988, at a total cost of about $120 million. Thankfully, the fires killed no park visitors and no nearby residents. Outside the park, two firefighters were killed, one by a falling tree and one while piloting a plane transporting other personnel.

Effects on Land, Property, and Wildlife

Ecosystemwide, about 1.2 million acres was scorched; 793,000 (about 36%) of the park's 2,221,800 acres were burned. Sixty-seven structures were destroyed, including 18 cabins used by employees and guests and one backcountry patrol cabin in Yellowstone. Estimated property damage totaled more than $3 million. About 665 miles of hand-cut fireline and 137 miles of bulldozer lines, including 32 miles in the park, needed some rehabilitation, along with the remnants of fire camps and helicopter-landing spots. Surveys found that 345 dead elk (of an estimated 40,000–50,000), 36 deer, 12 moose, 6 black bears, and 9 bison died in greater Yellowstone as a direct result of the fires; 2 radio-collared grizzly bears were missing and were presumed to have been killed, (although one turned up alive and well several years later). Most of the animals that died were trapped as fire quickly swept down two drainages, and were discovered when biologists subsequently observed scavenging grizzlies, coyotes, and birds feeding on the carcasses. A few small fish-kills occurred as a result of either heated water or dropping fire retardant on the streams. Surveys revealed that less than 1% of soils were heated enough to burn below-ground plant seeds and roots.

A massive effort was funded by the U.S. Congress to restore damaged facilities and to study the long-term ecological, social, and economic effects of the Yellowstone fires. Although the tourist season was cut prematurely short by the fires and associated firefighting activity, the feared abandonment of regional visitors failed to materialize in 1989. The effects on many plants and animals are still being studied, although in the short-term, most wildlife populations showed no effect or rebounded quickly from the fiery summer. In the several years following 1988, ample precipitation combined with the short-term effects of ash and nutrient influx to make for spectacular displays of wildflowers in burned areas. And, where serotinous lodgepole pines were burned, seed densities ranged from 50,000 to 1 million per acre, beginning a new cycle of forest growth under the blackened canopy above. . . .

In 1992, Yellowstone National Park again had a wildland fire management plan, but with stricter guidelines under which naturally occurring fires may be allowed to burn.

Although unprecedented in the 125-year history of the park, the scientists reviewing the effects of the 1988 fires reminded us that fires of such scale burned elsewhere in similar ecosystems during this century, and earlier in the landscape's history.

Trapped by California's 2003 Firestorms

BY SCOTT GLOVER, JACK LEONARD, AND MATT LAIT

In Southern California fire danger is most extreme in the fall. By September, the region has gone without rain for several months, and hot dry winds, called Santa Anas, begin to blow in from the desert. Added to these conditions in the autumn of 2003 was a multiyear drought that dried out chaparral and weakened millions of trees, which eventually succumbed to a bark-beetle infestation. Dry brush and dead trees—and houses built in rural areas—proved to be ideal fuel for wildfires.

Given these conditions, fire experts were not surprised when multiple wildfires began burning in several Southern California counties in October 2003. No one was prepared for the magnitude of the firestorms, however. Southern California's October 2003 fires were the worst fire disaster in California history, killing twenty-two people, destroying more than thirty-five hundred homes, and burning over 746,000 acres.

In the following selection, Los Angeles Times *staff writers Scott Glover, Jack Leonard, and Matt Lait describe the early hours of San Diego County's most destructive fire, which began in a remote area in the Cleveland National Forest. They explain how early Sunday morning, October 26, the Cedar fire reached residents living near the Barona Indian Reservation. With no warning from authorities, residents had only minutes to flee. Some made it to safety, but others died trying to save family members and neighbors. The Cedar fire killed at least twelve people, destroyed over two thousand homes, and charred over 275,000 acres.*

At least 12 people killed [in October 2003] in Southern California wildfires were from two rural San Diego County neighborhoods that survivors said received no warning from authorities.

When the Cedar fire reached residents here [near the Barona Indian Reservation] early Sunday [October 26], they had only minutes to flee. A dozen died—some in cars, some huddled in homes and others trying to run for safety.

Eight victims lived along a series of dirt roads just outside the Barona Reservation. Neighbors told *The Times* on Wednesday [October 29] that the dead included a 17-year-old high school student, her mother and the teenager's aunt, whose skeleton was found in a bathtub along with the bones of her dog.

"We feel really lucky," said Lonnie Bellante, who escaped with his wife and 11- and 13-year-old daughters. Four of his tenants and four of his neighbors died. "We got out with the most valuable thing, which is each other. The rest of it is not important."

Four people died in Lake View Hills Estates, a gated 10-home community a few miles west. Lakeside fire officials said they reached the entrance to the neighborhood about 3 A.M. Sunday, but were ordered to retreat before getting a chance to warn residents of the coming firestorm. Five homes also burned.

The fast-moving Cedar fire, sparked by a lost hunter Saturday afternoon [October 25], caught fire officials by surprise as it moved west about 15 miles from the Cleveland National Forest.

A spokesman for the reservation said the Barona fire station was notified of the fire about 2 A.M. Sunday by a California Department of Forestry battalion chief. He told the firefighter on duty that the fire would be passing through in 20 minutes, said Dave Baron, director of government affairs.

Baron said the fire station notified the reservation's casino and some nearby residents, but did not have time to notify Bellante and his neighbors.

Jon Smalldridge, who was spending the weekend at his parents' home near Bellante, said Wednesday that only the persistent barking of his dog gave him and his house guests enough time to escape the fire's path.

It also gave him a chance to save some of his neighbors.

Smalldridge, 42, of Arcadia, and his 18-year-old son, Shawn, had come to relax in country thick with trees and boulders. Smalldridge and his son spent Saturday riding motorcycles and target shooting. Jon Smalldridge's parents were away.

On Saturday evening, the Smalldridges watched the distant glow of the Cedar fire that had begun hours earlier in the Cleveland National Forest. Neither one worried, and they went to bed tired.

The dog's barking woke the household about 1:30 A.M. Sunday. Outside, the smoke was thick.

Smalldridge drove to neighboring homes to warn residents. One of them was Molly Sloan, a family friend and a grandmother, who didn't want to go. She had bought her 14-acre ranch 28 years ago. Her property had four houses—family members lived in three and renters occupied the fourth.

Smalldridge rounded up Sloan's three dogs—Sasha, a boxer; Freckles, a mutt; and Osa, a lab—and loaded them into Sloan's car. Then Smalldridge helped Molly Sloan into her car.

Mary Peace, Molly Sloan's daughter, followed in her own car. As they drove away, Mary Peace decided to check on neighbors and turned off the road.

Then Smalldridge, who owns an auto repair shop in Pasadena, and his son took off in their pickup. A car was stopped ahead. It was Molly Sloan's. Flames engulfed the road, 15 to 30 feet high, he said, whipped by 40-mph Santa Ana winds. Smalldridge slammed his foot on the brakes. He could not see inside Sloan's car.

"Dad, you gotta go. We gotta go. There is nothing you can do," Shawn told his father.

Smalldridge jammed his foot on the accelerator and sliced through a wall of flames. He couldn't see the road ahead. "I was driving by Braille," he said.

They reached a gas station by the Barona casino, and the firestorm surged past.

"We stayed in the car and watched the whole hillside go up," Smalldridge said. "Knowing there were still people who couldn't get out was pretty difficult. There was nothing more we could do."

After escaping the inferno, Jon Smalldridge at dawn agonized over whether to return to his neighborhood. He worried about Molly Sloan. But he also worried about exposing his son to a gruesome sight.

Smalldridge prayed. Then he turned the key in the ignition.

"I was mentally prepared for what I saw; I prepared myself to come back," Smalldridge said. "I . . . was willing to accept the worst. Knowing there wasn't going to be any medical personnel, I prepared myself to help them."

As he approached the cluster of homes, Smalldridge drove up to a charred Toyota. The left front wheel was in a ditch. A skeleton sat in the passenger seat, with the remains of a dog by its feet. Twenty feet away, Smalldridge found another body.

"I knew it wasn't going to be pretty," Shawn Smalldridge said. "I just prayed that it was quick for them." A downed power pole blocked the road, and the Smalldridges continued on foot. They saw a blackened car in a field. A woman's body lay sprawled 30 feet in front of the car.

It was Galen Blacklidge, who lived with her husband a few houses away from Jon Smalldridge's parents.

At 6 A.M., Smalldridge met up with Lonnie Ballante. Ballante, his wife and two daughters had followed the same route as Smalldridge, but the fire turned them back. The engine quit and they had to leave their truck. Ballante's wife suffered the worst burns, with the flesh coming off her arms. Smalldridge drove the family to get help.

Molly Sloan had survived, having followed Smalldridge. Her daughter, Mary Peace, did not. Peace had turned off the road to check on neighbors. Minutes later, her escape was apparently cut off by flames. The 54-year-old former nurse must have returned to her house, speculated Mike Parsons, a relative of Sloan.

Relatives later found Mary Peace's remains in a bathtub. With her were the remains of her Chihuahua.

Nearby, in another house on the Sloan family compound, they found the remains of 17-year-old Jenifer Sloan, Parsons said. Her mother, Robin, is missing and presumed dead. Robin Sloan had worked at a nearby Wal-Mart. She had driven to the trailer of her former husband to warn him, family and friends said.

He was already gone, but Robin stayed to pack mementos, they said. Then, they believe, she drove back up to her house.

Jenifer Sloan, an El Capitan High School student, remained waiting at the house.

As the fire bore down, Jenifer was on the phone trying to wake neighbors. She also called a friend who begged her to evacuate. She told the friend she would not leave without her mother.

"They all died trying to save someone's life other than their own," Parsons said. "They all had their own vehicles. They all could have left at any time. You might say it was a family of heroes."

Fire Prevention and Damage Control

The Fire Debate

By Timothy Egan

According to Timothy Egan in the following selection, in 1910 a huge conflagration in the West radically altered official U.S. Forest Service policy. Hoping to prevent such destructive fires in the future, Forest Service officials decided that from then on all wildfires would be fought. Egan explains that by midcentury, however, many fire experts began to doubt the wisdom of this policy. Because fires were not allowed to burn naturally, the West's forests became choked with brush and dead wood, ideal fuel for wildfires. As the century wore on, fires had increasingly more fuel to burn and quickly turned into uncontrollable conflagrations, destroying millions of acres and destroying thousands of houses. Egan reports that by the end of the century, the West's fire risk was even greater due to prolonged droughts and above-average summer temperatures.

Fire experts, environmentalists, politicians, and others are now engaged in a heated debate over how best to manage the nation's forests and other wild areas. As Egan explains, many fire experts contend that fires should be allowed to burn naturally. However, this argument has become ever more contentious as increasing numbers of people move into areas where the fire risk is high. Naturally, these property owners want all fires extinguished to protect their homes. On the other side of the debate are those who believe that the fire risk can best be reduced by increasing logging in the West's forests to reduce fuel levels. According to Egan, environmentalists strongly oppose such a policy, however, claiming that those who advocate increased logging are using wildfire risk as a scare tactic to gain support for a highly destructive commercial enterprise. Politicians, he reports, weigh in on both sides of the issue, depending on which stance will garner them the most votes. Timothy Egan writes for the New York Times.

T he fires came early [in 2002] to the West, chasing people out of valleys in Colorado, rousting animals from late slumber in Alaska, choking the sky with smoke in Ari-

Timothy Egan, "Era of the Big Fire Is Kindled at West's Doors," *New York Times*, June 23, 2002. Copyright © 2002 by The New York Times Company. Reproduced by permission.

zona woods that have so little moisture they seem kiln-dried.

The price of holding back nature has come home, fire officials say. A century-long policy of knocking down all fires has created fuel-filled forests that burn hotter and faster than ever. The era of big fires—and with it, the need for big government to contain them—is at hand, many firefighters say. Already, with 1.9 million acres burned by the first day of summer, wildfires across the West are burning twice the acreage of the 10-year average for this time of year.

A convergence of events—drier forests, higher temperatures, a yearslong drought and more people living in places where fire has long made a home—is likely to keep armies of yellow-shirted firefighters busier than ever, at a cost to taxpayers of $2 billion a year.

"We've got the equivalent of the perfect storm," said Stephen J. Pyne, an Arizona State University fire historian who has written many books on the subject.

Forest Service officials say 73 million acres, about 40 percent of all Forest Service land, are at risk of severe fires in coming years.

Since four firefighters choked to death on superheated gas in this [Winthrop] Eastern Washington valley [in 2001], government strategy for fighting fires has changed, with new rules that will slow response in the woods.

Smoke jumpers will not always be in Westerners' backyards at a moment's notice.

"Some citizens may resent any delay when they know there is a fire burning in the forest," said Sonny J. O'Neal, supervisor of the Okanogan and Wenatchee National Forests.

In Arizona [on June 23, 2002], two major fires that have been burning all week were about to connect, and officials told residents of the area's largest town, Show Low, to evacuate.

"We're at the mercy of Mother Nature right now," said Larry Humphrey, incident commander of the Rodeo fire, Arizona's largest. "There's not a whole lot we can do."

Some say the fires are a harbinger. "These catastrophic fire seasons are going to become the norm," said Bruce Babbitt, the former Interior secretary and Arizona governor. "The question is, what are we going to do about it? Can we learn to live in the woods, when in most of these areas there aren't even building codes?"

One central question is whether the government should be more willing to start controlled fires, to burn off built-up fuels. But the policy is vexed, in part because some of the biggest recent fires were government-started blazes that got out of control—and because growing numbers of people and homes are in harm's way if controlled burns jump the rails.

The fires this time are also prompting calls to enact a new social contract. People living in fire zones would have to do preventive maintenance to expect government help when the woods catch fire. The insurance industry, which has forced a change among home developers by making it more costly to live in flood zones, is considering similar rules for fire areas.

In Alaska, where a half-million acres have burned [in 2002] and the fire season came earlier than anyone can remember, insurers have already stopped offering policies to homeowners who refuse to remove fire hazards from their houses in areas where dead spruce trees are likely to burn.

But changing fire policy is slow, subject to partisan fluctuations and interest-group pressure and fraught with technical questions. "Reinstating fire is like reinstating a lost species," Dr. Pyne said. Republicans blame environmentalists, arguing that stepped-up logging is the answer, to clear the forests of the trees most likely to burn. Democrats argue that the timber industry is using fire as an excuse to cut down trees.

The Bush administration has no plans to change fire policy, said Mark E. Rey, who oversees the Forest Service. It will try to reduce the "process paralysis" that has kept land agencies from taking big new steps to clear out fuel in the forests, he said, and to encourage people in fire zones to be aware of the constant threat.

On the fire front lines in Colorado and Arizona, the message is starting to get through. But for many, it is too late.

The Urban-Wildland Interface

After deciding to leave Las Vegas, Nancy and Steve Smith looked at nearly 40 houses before finding the 3,200-square-foot home they bought [in March 2002] in the Colorado Rockies.

"We knew this was the house we had been looking for," Mrs. Smith said. "It was on top of a wooded ridge overlooking a valley with horse ranches. To the west we had beautiful views of Thunder Butte. Our property bordered Pike National Forest on one side."

They came to the mountains, like most urban exiles, seeking clean air, solitude, a closeness to nature.

"I had no idea until mid-April that we were in an area that was considered high risk for fire," Mrs. Smith said. "When we moved in, there was snow on the ground. And it snowed every week for a month."

On May 19, barely a few weeks after that last snow, the Smith family was told to evacuate because of a growing wildfire. Then came the Hayman wildfire, Colorado's biggest blaze, which the authorities say was started by a seasonal Forest Service worker. In days, the fire covered 20 miles, spreading to within five miles of the Smiths' house.

Last Monday [in June 2002], the Smiths had two hours to retrieve more possessions. On Wednesday, their dream house was destroyed.

Whether the fire was natural or not will be debated for years. But the Smiths' story is more common across the West as more people move into what used to be wilderness.

Experts call the zone where homes meet forest the "urban-wildland interface." That is where most fires are being fought.

"Ten times as many homes are now in areas prone to wildfire as there were 25 years ago," said Don Smurthwaite, a spokesman for the National Interagency Fire Center in Boise, Idaho.

In Colorado, the number of people living in areas at risk of fire increased 30 percent during the 1990's, said James E. Hubbard, the state forester. Nearly a million people in Colorado now live in the fire zone, state and federal demographers said.

Demography and nature collided in Colorado [in spring 2002] when wildfires broke out in the Rockies, where a drought meant there was little snowpack to provide moisture.

"Usually, the fires don't start until June in this part of the world," said Gordon Koenig, a pilot who has been dumping fire retardant on Colorado six days a week since April. Mr. Koenig said the season had doubled in length in the 13 years he has been fighting fires.

A half-century ago, the strategy of fighting nearly every fire was easier. But the era when smoke jumpers would drop into uninhabited valleys is gone, Forest Service officials say. Its end is complicated because the new generation of Westerners has not learned how to live in the red zone safely.

"While Westerners are getting more educated about fire, there is still a kind of dangerous independence, one that resists all zoning and regulation, that exists among people who live in the fire zone," said Pat Williams, a former Montana congressman.

Others agree. "I think the public has accepted the fact that fire has a natural place," said Dr. Pyne, the Arizona State fire historian. "People are starting to get it. But it will take another 10 years or so to work out. Unfortunately, that means the peak of wildfires and homes destroyed will be in the next five years."

Fire Management: A Burning Issue

Smokey Bear came of age in an era that was haunted by a single summer, 1910, a year that still hangs over all Forest Service decisions about fire. In that year, it seemed as if all of the West was on fire. Three million acres burned in Montana and Idaho alone, and 87 people died. "Thousands of people thought the world was coming to an end," wrote Norman Maclean, the author of "Young Men and Fire," a story of one of the worst calamities in firefighting history.

The deaths, and the apocalyptic images of valleys where daylight had turned to darkness, prompted the fledgling group of government foresters to adopt a new policy. From then on, every fire would be fought, quickly.

But even by midcentury, some foresters were beginning to argue that the policy was misguided and that by snuffing out all fires, the Forest Service was only delaying the inevitable big fires.

A 1999 report by the General Accounting Office blamed a century of putting out all fires for "an increasing number of large, intense, uncontrollable and catastrophically destructive wildfires." The agency said forests in the West would be at risk of big fires through 2015.

The government has made various attempts to change its policy. One of its most notable experiments was in 1988, when some natural fires that consumed more than half of Yellowstone National Park were allowed to burn.

Though the policy angered some Western senators at the time, the now-green park seems to vindicate the National Park Service's decision.

But the issue of allowing fires to burn becomes infinitely more complex when the fire is raging not in a park or a wilderness but near a neighborhood.

"People will simply not tolerate that," Mr. Babbitt said.

The government's ability to make the case for controlled burns has also been severely hampered by its own missteps. In 2000, for example, a fire started by the Park Service in New Mexico raged out of control and destroyed hundreds of houses and thousands of acres, ultimately costing the government far more than it ever would have paid to fight a natural fire in the area.

Fires become politicized, too. In 1988, some people blamed the Reagan administration for letting Yellowstone burn. In 2000, George W. Bush implied that President Bill Clinton's policies were to blame for Montana fires.

"The only thing that burns hotter than a wildfire in the West is the demagoguery of some politicians trying to take advantage of it," said Mr. Williams, the former Montana congressman.

The Clinton and Bush administrations have pushed for some logging of dead or dying forests in particularly vulnerable areas, but have been stymied by lawsuits and protests from environmental groups.

[On June 16, 2002], in testimony before Congress, Dale Bosworth, the chief of the Forest Service, said "analysis paralysis" from lawsuits and second-guessing by the land agencies had prevented the government from burning or logging some fire-prone areas.

But even in areas that have been logged, laced with roads or cleaned of excess brush, firestorms have raced through. In 2000, when eight million acres burned, fires scorched the Bitterroot Valley in Montana, taking out ancient trees on one side of the mountains and homes and orchards on the other.

The other policy debate centers on firefighters and how much risk they should take to save property. When the Mann Gulch fire killed 12 young men in 1949, the Forest Service vowed to never repeat its mistakes.

But in 1994 in Colorado, 14 firefighters died in the South Canyon fire. The circumstances were hauntingly similar, except this time they were fighting to save houses in the urban interface. The Forest Service partly blamed gung-ho firefighters for the deaths.

[In 2001], after Congress met the call for more firefighters in response to the huge fires of 2000, four people died in the Thirty Mile Fire in the valley just north of here [Winthrop, Washington State], trapped in fire shelters after a flame storm overwhelmed them. An investigation said numerous safety rules had been violated. Some people also blamed the rush to throw firefighters, some poorly trained, at fires.

All these deaths have had a humbling effect on the men and women who fight fires. As Mr. O'Neal, the forest supervisor, said in announcing new guidelines . . . , firefighters are not going to die to save property.

"We will continue to attack and control fires that threaten life, property or important natural resources," Mr. O'Neal said. "But in every case, safety comes first.". . .

Learning to Live with Fire

Like other Western states where dry conditions and dense underbrush are fueling blazes, Colorado is learning a hard lesson.

It was a lesson residents of Malibu, California, learned in 1993, when a firestorm wiped out more than 350 homes.

After the blaze, wholesale changes were made in building codes, and even in the rules governing landscaping, to reduce the spread of fire. The changes are among the most stringent in the country. . . .

It makes a difference. Though fire officials are expecting one of the worst fire seasons ever in Southern California because of a drought and a heavy accumulation of brush, they also say that the lessons of 1993 have helped to reduce the risks. . . .

[Kathy and John] Haag . . . had just finished a two-year renovation of their home when the 1993 fire roared through. They escaped with only a few things thrown into the back of a car. Seeking to expand their home later, they ran into an array of tough new standards.

They have to leave a five-foot space around the house clear for access. Grass has to be mowed to three inches or less, and ground cover, which can start 20 feet from the house, must be 18 inches or lower. Trees must be 30 feet apart. Eaves must be covered, with few ventilation openings, to prevent embers from lodging there in a fire. New houses must have sprinkler systems. . . .

It remains to be seen whether communities in Colorado and throughout the West, not all as rich as Malibu, will follow its example.

Some experts are not optimistic. "We call ourselves a nation of pragmatists," said Dr. Pyne, the fire historian, who spends much of the year living in the fire-prone woods near Alpine, Arizona. "But you wouldn't know it by the way we deal with fire."

Predicting Wildfires with Computer Simulations

By Patrick L. Barry

One of the most challenging jobs of fire managers is to choose the right strategy to prevent fires. The right strategy will help control and suppress fires while the wrong strategy could create conditions ripe for a destructive firestorm. To prevent wildfires experts have to consider many factors pertaining to the region, including the amount and type of vegetation and its moisture content, topography, weather, and ecological data. In this selection Patrick L. Barry explains how experts are now inputting this kind of information into computer simulations to help them determine the best fire prevention strategies. For example, experts may use the simulation to determine how different amounts of vegetation would burn if they caught fire. Fire managers then know how much vegetation to clear to reduce fire risk. Patrick L. Barry is a writer for Science@NASA, an online news service provided by the U.S. National Aeronautics and Space Administration.

It begins with a quick flash of blue lightning. The weather in this remote wooded valley is stormy, but rain hasn't fallen on the parched forest in weeks.

Fanned by the high winds, the small flame sparked by the lightning spreads quickly in the thick mat of dried twigs, pine needles, and fallen branches. Standing dead trees catch, and within hours the blaze has lit the canopy. A modest spark is quickly escalating into a blistering firestorm.

Patrick L. Barry, "Fighting Wildfires Before They Start," http://science.nasa.gov, August 28, 2001.

"Hmmm, that's no good," says the forest manager as she regards the computer screen alongside her colleague. "Let's try it again. But this time," she muses, "let's thin out the dead trees and trim all the lowest branches up to 8 meters from the ground."

The scientist makes the changes with just a few clicks of the mouse, then restarts the fire with another bolt of virtual lightning. This time the fire just creeps along the ground, clearing out the excess clutter, and sparing the adult trees.

This scenario might sound futuristic, but computer simulations of forest fires are already transforming how land managers protect their forests and the people who live near them. By combining satellite-derived vegetation data with topographic maps, weather data, and ecological knowledge, forest scientists can construct digital landscapes on which these virtual fires burn.

Computer Simulations: A New Way to Fight Fires

The computer-assisted approach to fire risk assessment is still relatively new and only partially adopted by the fire management community, but the advantages of using computers have become widely recognized, and the technology is spreading like—well, like wildfire.

"It's well recognized now that this is what needs to be done," says Mark Finney, a forest researcher with the USDA [U.S. Department of Agriculture] Missoula Fire Sciences Laboratory in Missoula, Montana.

"Having a digital map of forest characteristics and simulating the fire behavior of the whole map in a computer is really the future of planning efforts," Finney says.

As a ballpark estimate, Finney suspects that only 15 percent of the forestry community utilized these high-tech tools just 2 years ago. Now, he estimates, that figure would be closer to 40 percent, and within 5 years he expects the technology to be nearly pervasive.

Simulating the spread and intensity of a wildfire clearly has important uses for officials who decide how best to battle an ongoing blaze. But perhaps an equally important use of these computer models is to aid decisions about how to reduce the risk from forest fires before they happen.

Forest managers have a few tricks up their sleeves—called "fuel treatments"—that allow them to lessen the chance that a poten-

Firefighters clear dry brush from the perimeter of a home in an attempt to save it from a wildfire.

tial forest fire will be dangerous. Low-intensity fires are a normal part of many forest ecosystems. They clear underbrush, open seed pods, and return nutrients to the soil. Mature trees normally survive these "ground fires." But if an area contains too much dead wood and dry leaves—that is, fuel—the tops of the trees can be ignited, starting a "crown fire" that kills everything in its path.

The solution is often to remove some fuel from the area. There are several ways to do this, such as controlled burns, selective logging, or trimming low branches and underbrush, to name a few. By using the computer models, managers don't have to merely guess at the best choice. They can [create] simulations of each option and compare the results.

Often the effects of an alteration can be complex and counterintuitive.

"We can evaluate the effectiveness of a treatment over time, because we can simulate the regrowth of the forest. Treatment *A* might be more effective for reducing the spread of a fire for the first 10 years, but treatment *B* might be more effective than *A* for years 10 to 50," says Kevin Ryan, project leader for the Fire Effects Project, also at the Fire Sciences Laboratory.

In the case of prescribed [controlled] burns, these simulations can be particularly important.

"Fire managers are quite reluctant to set prescribed fires if they can't determine how those fires will spread," says Steve Running, a professor of forest ecology at the University of Montana who specializes in remote sensing applications. "In the Los Alamos fire [May 2000], the big problem was that fire managers guessed how the fire would spread, and they guessed wrong, so the fire burst out of control."

Creating Virtual Landscape with Satellite Images

In order for the models to produce predictions similar to the behavior of a real fire, scientists must provide the computers with "virtual landscapes" sufficiently similar to the real terrain. With millions of acres to map, scientists must rely on remote-sensing satellites. The workhorse for land mapping in recent years has been Landsat 7 [short for "land satellite"], which was prepared for operation and launched by NASA's [National Aeronautics and Space Administration's] Goddard Space Flight Center [GSFC], then turned over to the U.S. Geological Survey [USGS] for daily operations. The 30 m-resolution maps produced by Landsat lay the foundation for the vegetation data fed to the models.

"Data from the Landsat imager can be used to produce vegetation index maps," says Darrel Williams, a forest ecologist and the Landsat project scientist at GSFC. "The vegetation index for a given area can vary from season to season and from year to year, especially for grasslands and scrublands. One can assess if there is a greater fuel load than normal." That can be a danger sign if conditions become drier.

Landsat maps reveal the boundaries between forests, grasslands, deserts and cities. But that's not enough. To a wildfire, not all forests are created equal. The virtual terrain needs to include more detail about the forest ecosystems than 30m pixels alone can provide.

Land managers need to know what is the dominant tree species? How dense is the canopy? Do the trees range in age, or is the stand more uniform? When was the last fire in the area? Some of these questions can be answered by inspecting the history of the area's vegetation. For others, scientists must rely on their extensive ecological knowledge and on fieldwork to find clever ways to infer such details from the satellite data.

"If two polygons on the map have similar spectral images, by knowing at what slope aspect and elevation and latitude that polygon is, our ecological knowledge tells us, well, it can't really be spruce, it has to be Douglas fir, for example," Ryan says.

This contour information is deduced by laying the Landsat data over the top of a digital topographic map produced by the USGS, creating a 3-dimensional landscape. Artificial structures like roads and buildings are also added, which allow the forest managers to see how a potential fire would affect the local communities.

The Weather Factor

Finally, the crucial ingredient is added: weather. Moisture and wind can make or break a forest fire. Since moisture lingers on the terrain, historical records of the area's weather must be woven into the model along with current conditions.

Like a forester's *SimCity 3000™* [a role-playing computer game], the computer software integrates the vegetation, the terrain, and the weather into a "virtual forest" that simulates the interactions of all these variables.

Trying out fuel treatments *in silico* holds great promise for helping forest managers mitigate fire risk, but the technology has only been in routine use for about 5 years and still has much room for improvement.

"It's an inexact and emerging science, obviously," Ryan says. "There still is a lot of research and development going on in fire behavior modeling and the effects of the different treatments."

New satellite data should also bring some improvements to the field. For example, researchers are learning how to better estimate forest canopy structure using the Multi-angle Imaging Spectro-Radiometer (MISR) on NASA's Terra satellite, launched in 1999. MISR provides 9 different views of the terrain below, each from a different angle.

"If you just look straight down at the trees [it is] often hard to detect how thick the canopy is. But if you can look from various directions, the 3-dimensional nature of that canopy is much more clear," Running says. Canopy structure is an important factor, since it partly determines whether a ground fire will escalate into a raging crown fire.

In the future, this marriage of computer and satellite technologies should become a robust tool for helping reduce the risk posed by wildfires.

"As computers get faster and models get better, then prediction will get better, too," Finney says.

Smoke Jumpers

By Glenn Hodges

In this selection journalist Glenn Hodges relates his personal experiences with Russian smoke jumpers during the summer of 2002. He explains that smoke jumpers are a group of firefighters trained to fight fires in wild areas that are hard to reach by ground transportation. To reach a fire, two or more smoke jumpers parachute from an airplane or rappel from a helicopter, carrying their tools, food, and survival supplies. According to Hodges, although Russian and American smoke jumpers use many of the same techniques to fight fires, such as clearing brush away in front of a blaze and then lighting a back fire, Russian smoke jumpers have to do their job with inferior equipment. Russia, for example, cannot afford to supply firefighters with fire-resistant clothing. To compensate for the lack of equipment, Russian smoke jumpers have become experts in developing survival and firefighting skills using few resources, Hodges reports. Smoke jumpers have one of the most dangerous jobs in Russia, but they all agree that they would not exchange smoke jumping for any other job in the world. Glenn Hodges is a staff journalist for National Geographic *magazine.*

A lexander Selin, the head of central Siberia's aerial firefighting force, is a man who knows how to make himself clear, even in English, a language he barely knows. . . .
After a few days in his care we will come to call Alex, simply, Big Boss. A thick-fingered, barrel-chested Siberian who hurls his words like shot-put balls, Alex rules a fiefdom the size of Texas with an army not much bigger than the Texas A&M marching band. His 500 smokejumpers, firefighters who jump from planes and rappel from helicopters, cover a swath of boreal forest that stretches from Arctic tundra to the Mongolian border.

Photographer Mark Thiessen and I have come to Siberia to see Alex's men in action, but by the time we're halfway from Krasnoyarsk to Shushenskoye, 200 miles to the south, I'm not sure we'll live long enough to see a single fire. We're throttling through the mountains in a pair of fume-filled Volgas, taking curves at 90 miles an hour, passing blind on hillcrests, narrowly avoiding one head-on collision after another—and I'm thinking wistfully back to our training in British Columbia, where Mark and I rappelled out of helicopters feeling as safe as the day we were born. . . .

So I'm not surprised the next morning when we board our first Mi-8, an 18-wheeler of a helicopter that is Russia's aerial firefighting workhorse, and there are no seat belts in sight—and practically no seats. Alex has taken our visit as an opportunity to host a half dozen cronies on a weekend fishing trip in the mountains, and when we land in a field to pick them up, gear gets piled willy-nilly between the two huge fuel tanks—a rubber boat here, an outboard motor there—and everyone plops down on whatever looks most comfortable.

That afternoon, over vodka shots at the fishing camp, Alex explains the Russian way of doing things. He's been to California and Idaho to see how American firefighters work, and when he thinks of riding in their helicopters—all strapped in by seat belts and regulation—he laughs at the memory. "No move, no speak!" he says. You can't size up a fire if you can't move around and look at it! You can't make a plan if everyone has to be quiet!

"And they call Russians crazy!" the pilot cuts in.

Having barely survived their driving, I'd say "crazy" seems about right, but you've got to be at least a little crazy to jump from a plane into a fire, and the Russians have been doing it longer than anybody.

The First Smokejumpers

"The idea of actually parachuting into fires was a Soviet invention," I am later told by Stephen Pyne, an American wildfire historian who is one of the few people outside Russia who know much about Avialesookhrana, Russia's aerial firefighting organization. "In the 1930s these guys would climb out onto the wing of a plane, jump off, land in the nearest village, and rally the villagers to go fight the fire."

[In 2001] Avialesookhrana celebrated the 70th anniversary of its first flight. (It would have been the 75th anniversary, but when the first plane took off from Leningrad in 1926 to look for fires, the pilot made a beeline for Estonia and defected.) Once they got going, the Soviets quickly built a program that remains to this day the largest in the world—despite a decade of post-Soviet budget cuts that have halved the ranks, from 8,000 firefighters to 4,000. It's a shoestring operation—just $32 million a year to cover 11 time zones, less than the United States might spend in a few days of a heavy wildfire season. But with their mismatched uniforms and 50-year-old biplanes, Russian smokejumpers do what their countrymen do so well: make do with less. Less money, less equipment, and yes, less caution—even with fire.

When we break camp the next day to return to Shushenskoye, I'm surprised to see that the campfire is left smoldering. It's a hot July day, which would be bad enough without the helicopter's rotor wash blowing everything all over the place, but the risk doesn't even seem to register with Alex, central Siberia's most powerful firefighting official. In the U.S., firefighters would douse a fire on an ice floe in the dead of winter, especially with journalists around. But here they play the odds the way they see them, and perfect safety is burdensome and unnecessary. Fire shelters and fireproof clothing? Too expensive, but that's OK, because the odds of needing them are low. Seat belts? Impractical. Thousands of times you will buckle and unbuckle, and probably for nothing. Campfire? It's not going anywhere.

Not surprisingly, people cause two-thirds of Russia's 20,000 to 35,000 annual wildfires—and by the time we've been in Siberia for a week, I find myself wishing they'd cause a few more. The Shushenskoye area is hot and dry but fireless, and after seemingly endless rounds of vodka-steeped hospitality we finally persuade Alex to send us north to Yeniseysk, where we hear fires are burning across the region.

Fighting Fires

When we get to the base in the town of Yeniseysk two nights later with our guides, firefighters Valeriy Korotkov and Vladimir Drobakhin, we're eager to finally get to a fire, and the gods respond accordingly: In the morning it is raining. Pouring. I look at Valeriy. "I thought you said Friday the 13th was your lucky day."

"Ahh, the day is not over, my friend."...

Midday the word comes: Hurry and get your stuff together, we're going to a fire. I'm just this side of incredulous, given that we've been socked in by rain for 12 hours, but a two-hour helicopter ride later, we land at the edge of a smoldering patch of forest and the sun is shining. Valeriy's lucky day! He and Vladimir quickly chop down some birch saplings to make poles for our canvas tent, and we hike what looks like an old logging road through a blackened forest to the fire line.

Twelve smokejumpers have been on this 120-acre fire for nearly a week. It seems all but dead on this flank, but the guys chop a few saplings to make handles for their rakes and shovel blades and get to work, scraping clean a foot-wide swath of forest floor, then lighting a backfire with pine needles and birch bark. The backfire burns toward the wildfire, consumes its fuel, and stops it in its tracks—the basic technique of wildfire fighting everywhere, whether done with shovels and pine needles or bulldozers and drip torches.

Each summer Avialesookhrana's firefighters face the Herculean task of containing fires across two billion acres of the largest coniferous forest in the world. Though regional forestry offices help fight fires in more populated areas, smokejumpers—housed in 340 bases across the country—are the sole defense for half of Russia's territory, flying to fires in crews of five or six when parachuting from An-2 biplanes and in groups of up to 20 when rappelling from the Mi-8 helicopters.

"We face danger three times: one when we fly on plane; two when we jump; three when we go to fire," Valeriy says, and the statistics bear him out. In the past three decades 40 Avialesookhrana firefighters have died on the job—24 while fighting fires, 11 while parachuting, four in aircraft accidents, and one by lightning. Valeriy and Vladimir both tell me stories of parachuting fatalities, one when a jumper landed in water and drowned, another when a jumper hit an electric line. But jumping is the thrill that gets them hooked. "Two minutes fly like eagle, three days dig like mole," Valeriy says of the smokejumper's life—and the flying's worth the digging.

The day is late, so after making a couple hundred feet of fire line the guys break for a smoke.... As we swap Russian and English swear words and laugh, Alexi Tishin, an earnest 28-year-old with a week of stubble and a smattering of gold teeth, says, "This is the best job for tough guys"—you get to jump out of planes, fight fires, live in the forest. He says he especially loves jumping to small fires and trying to put them out fast. If they kill the fire in a day or two, they get a few extra dollars each—no small sum, given that smokejumpers earn on average 3,100 rubles, about a hundred dollars, a month. The incentive seems to work: More than half of all fires are put out within two days.

Russian and American Smokejumpers

The smokejumpers are true woodsmen—hunting, fishing, and trapping sable in the off-season to make ends meet, as nimble with an ax and knife as they are with their hands. When they land at a fire and make camp, they don't just make tent poles and shovel handles from saplings, they make tables, benches, shelves— you name it. I'm amazed to see one guy make a watertight mug out of birch bark.

It's a good thing their outdoor skills are solid, because their equipment often isn't. When we return from the fire line, Valeriy discovers that one of his brand-new experimental smokejumper boots has melted. The rubber sole is a mash of black goo. His boots lasted "an hour, at best" he says angrily, before launching into a torrent of complaint about poor Russian equipment. "This tent like from Second World War," he says, pointing at the canvas tent that will welcome mosquitoes and rain into our lives for days to come. The tents have no mosquito netting, the chain saws are heavy and unwieldy, the backpacks have no waist straps, the pull-on boots are made of cheap synthetic leather (and feet must be wrapped in towels to make them fit), the clothing is neither fire retardant nor water resistant. And everything is heavy.

For most of these guys that's just the way it is, but Valeriy and Vladimir are among the 120 Russian firefighters and managers who have been to the United States through an exchange program that began in 1993 between Avialesookhrana and the U.S. Forest Service. American and Russian exchangers alike are struck by the Americans' superior equipment and the Russians' inimitable resourcefulness.

Vladimir, who fills his American fireproof clothing with the stocky build of a linebacker, came home from a summer in the States with boots and tools and a wad of cash that was many times his annual pay, but he also returned with a new appreciation of his Russian brothers. "Put us in the woods with matches and a fishing rod, and we can live," Vladimir says. "We know how to eat mushrooms, catch fish, make a snare for animals. But for American firefighters, it would be a very bad situation."

Valeriy tells me how one time his squad's food was lost when it landed in the middle of a lake. They didn't have fishing gear, so he made a fishhook from a piece of metal on his reserve chute, pulled string from his parachute bag, cut a birch branch, and—voilà—they had fish.

By morning the rain we escaped in Yeniseysk has caught up with us, and we huddle under a tarp as we listen to the daily radio dispatch. A group of firefighters is stuck in the forest 200 miles to the northwest. There's no fuel to either fly them out or fly food in, so the dispatcher suggests that they build a raft and float down the river. No, they say, there's no good wood here for a raft. Then you'll have to walk, they're told—12 to 15 miles out, with all their heavy gear. You can almost hear the groans.

Fuel—or the lack of it—is a perennial problem for Avialesookhrana, even more the firefighter's bane than lousy equipment. Because we have a short time to see firefighters at work, Mark and I have been getting special treatment with helicopter transportation. But at the next fire, our luck runs out, and we get a taste of what smokejumpers have to put up with.

A Smokejumper's Day

The first night there we're drenched by the same weather system we've been fleeing since we got to the Yeniseysk region. After two nights and a day of rain, the sky clears, and we brave the mosquito hordes to dry our stuff and wait in vain for the helicopter. The day after that, still no helicopter. We're later told that someone back at the base forgot to fill out the correct paperwork—and the camp's radio battery has died, so we can't even call in a reminder.

"Every time we have same problem," Valeriy says. "After rain, they think, 'Guys sit in forest? It's OK.'" Once he had to wait 15 days for a pickup.

When the helicopter finally comes, we've been in Siberia for nearly three weeks, we've seen a grand total of 45 minutes of very sleepy fire, and we're told that it snowed two inches in Yeniseysk that morning. It's the middle of July [2002] and it has been the wettest fire season here in 15 years. We decide to leave for Vladimir's region in northwest Russia, where it's hot and dry and fires are breaking out all over the place.

Vladimir's base in Syktyvkar, a city of 226,000 roughly 600 miles northeast of Moscow, is much like the one in Yeniseysk—a central building with offices and training facilities, plus a dormitory where the smokejumpers live during fire season, from late

spring to early fall. The forest in this region is much more populated than what we've seen in Siberia, spotted with clear-cut
logging operations that make it easier for local foresters to get to
fires with bulldozers and local manpower.

As we fly over the checkered terrain in an Mi-8, we pass a
number of smoke plumes before landing near a square-mile fire
that horseshoes around a boggy meadow. A crew of five smokejumpers is already camped in the middle of the meadow, and less
than a mile away local forestry folks are cutting line in the forest
with bulldozers. Shortly after we land, an An-2 biplane circles the
fire and drops a hand-drawn map into our camp, and we hike
west through the woods to stop one edge of the fire.

The fire is a slowly moving wall of flame a foot or two high,
occasionally crowning into brush and treetops in quick whooshes
of flame. The guys light a backfire with pieces of birch bark, and
as the backfire burns forward to meet the fire, they extinguish its
back edge with devices called "piss pumps"—shoulder-strapped
rubber bladders that spray water through a nozzle. The backfire
itself isn't even necessary after a point, and they take to knocking down the flames of the wildfire with spruce boughs,
Vladimir leading the way. I catch the bug and see how much of
the fire I can contain just by stomping on it with my boots. I take
out a good 30 feet in a few minutes, and it's surprisingly satisfy-
ing. I have changed the course of nature.

Valeriy smiles at me and nods. Now I understand his job. "This
is the best!" he says. "We work in this forest no for money. We
work for our happiness. Not like people in Moscow.". . .

I ask Vladimir whether he thinks all fires need to be fought,
or if some should be allowed to take their natural course. "Fires
are natural, but bosses don't understand about letting fires burn,"
Vladimir says. He screws his face to one side and puffs out his
chest like he always does when he imitates bosses. "They say,
'Every fire we have to put out, because it's dangerous.'"

In truth, they don't put out every fire, but that's only because
they can't. In Siberia's remote expanses, where fire control would
be prohibitively expensive, fires are allowed to burn. "Not being
able to reach all these fires is probably for the good," says wildfire expert Stephen Pyne. "Fire is very much a part of the boreal
ecosystem."

Ten of the Worst Urban Fires in History

Date: September 2, 1666
Place: London, England
Dead: unknown
Damage: Many buildings, homes, and Saint Paul's Cathedral worth £10 million were destroyed.

Date: December 16, 1835
Place: New York City
Dead: unknown
Damage: Five hundred thirty buildings were destroyed.

Date: October 8, 1871
Place: Chicago, Illinois
Dead: about 300
Damage: More than seventeen thousand buildings were burned, costing about $190 million.

Date: November 9, 1872
Place: Boston, Massachusetts
Dead: unknown
Damage: Eighteen hundred buildings were destroyed, costing about $75 million.

Date: December 8, 1881
Place: Vienna, Austria
Dead: 620
Damage: Ring Theatre was destroyed.

Date: March 10, 1906
Place: coal mine in Courrières, France
Dead: 1,060
Damage: A mine was destroyed.

Date: December 7, 1946
Place: Winecoff Hotel, Atlanta, Georgia
Dead: 119
Damage: Most of the hotel was destroyed.

Date: April 16, 1947
Place: Texas City, Texas
Dead: 516 (more than 3,000 injured)
Damage: A fire destroyed the city and caused the explosion of the French freighter *Grandcamp.*

Date: May 13, 1972
Place: Osaka, Japan
Dead: 118
Damage: The fire destroyed a nightclub on top of a Sennichi department store.

Date: December 25, 2002
Place: shopping center in Luoyangi, China
Dead: 309
Damage: Disco locale and part of the shopping center were destroyed as a result of a Christmas party at an unlicensed disco.

Ten of the Worst U.S. Forest Fire Disasters

Date: October 8, 1871
Place: Peshtigo, Wisconsin
Dead: about 2,000
Damage: More than 2,000 square miles were burned; the town of Peshtigo was destroyed.

Date: September 1, 1894
Place: Minnesota
Dead: 600
Damage: Two hundred fifty square miles were burned; the fire destroyed six towns, including the town of Hinckley, where most deaths occurred.

Date: August 10, 1910
Place: Idaho and Montana
Dead: 85

Damage: The Big Blow Up burned more than 4,000 square miles of forests. After this fire the Forest Service decided to apply the put-out-all-fires strategy to prevent future disasters.

Date: August 5, 1949
Place: Mann Gulch in Helena National Forest, Montana
Dead: Thirteen of the sixteen smoke jumpers on the scene died when a fast-running fire caught them.
Damage: 6.8 square miles burned

Date: August 1988
Place: Yellowstone National Park, Wyoming
Dead: 0
Damage: More than 1,800 square miles were burned.

Date: October 20, 1991
Place: Oakland/Berkeley, California
Dead: 24
Damage: A brush fire burned more than three thousand homes and apartments located close to the wildland.

Date: July 2, 1994
Place: South Canyon, Colorado (also called the Storm King Fire)
Dead: Fourteen firefighters who were caught in a massive firestorm
Damage: About 3 square miles were burned.

Date: April 2000
Place: northern New Mexico
Dead: 0
Damage: A prescribed fire started by the National Park Service raged out of control and forced the evacuation of more than twenty thousand people. The fire burned more than 73 square miles, destroyed 235 structures, and threatened Los Alamos National Laboratory.

Date: Summer 2000
Place: western United States, including Alaska, Idaho, Wyoming, Montana, New Mexico, Nevada, Oregon, California, and Washington
Dead: undetermined

Damage: This was the most destructive fire season in U.S. history. Almost 12,500 square miles were burned, nearly double the national ten-year average.

Date: Summer 2002
Place: Hayman Fire in Pike National Forest, Colorado
Dead: undetermined
Damage: This was the worst wildfire in Colorado history; it burned more than 215 square miles and six hundred structures.

Date: October 2003
Place: southern California
Dead: 22
Damage: 1,172 square miles of land burned; 3,640 homes, 33 commercial properties, and 1,141 other structures destroyed (as of December 1, 2003). This combination of fifteen large fires was the worst wildfire in California history.

backfire: A fire set along the inner edge of a fire line to consume the fuel in the path of a wildfire and/or change the direction of the main fire.

combustion: The chemical process of burning, called oxidation, that produces new chemicals, heat, and light.

conduction: The transmission of heat between objects that are in direct contact with each other.

conflagration: A large, destructive fire.

convection: The transfer of heat by air currents.

crown fire: The movement of fire through the crowns of trees.

fire front: The part of the fire where continuous flaming combustion takes place. It is usually the leading edge of the fire.

fire line: A linear fire barrier that is scraped or dug down to bare (mineral) soil.

fire triangle: Graphic representation of the association between the three factors (oxygen, heat, and fuel) necessary for combustion.

fuel: Any combustible material.

oxidation: A chemical reaction in which a substance combines with oxygen.

prescribed burning: A planned fire to improve and restore a natural environment or to prevent massive wildfires.

smoke jumper: A firefighter especially trained to fight wildfires located in remote areas by parachuting from an airplane.

spotting: Behavior of fire that produces sparks or embers that, carried by the wind, start new fires beyond the original fire.

wildfire: An unplanned fire in a natural area including grass fires, forest fires, and other vegetation fires. It may be caused by people or be natural in origin.

FOR FURTHER RESEARCH

Books

David Cowan, *Great Chicago Fires: Historic Blazes That Shaped the City.* Chicago: Lake Claremont Press, 2001.

Frank Field and John Morse, *Dr. Frank Field's Get Out Alive: Save Your Family's Life with Fire Survival Techniques.* New York: Random House, 1992.

Margaret Fuller, *Forest Fires: An Introduction to Wildland Fire Behavior, Management, Firefighting, and Prevention.* New York: John Wiley, 1991.

Denise Gess and William Lutz, *Firestorm at Peshtigo: A Town, Its People, and the Deadliest Fire in American History.* New York: Henry Holt, 2002.

Jack Gottschalk, *Firefighting.* New York: DK, 2002.

John E.N. Hearsey, *London and the Great Fire.* London: John Murray, 1965.

Peter M. Leschak, *Ghosts of the Fireground: Echoes of the Great Peshtigo Fire and the Calling of a Wildland Firefighter.* San Francisco: HarperSanFrancisco, 2002.

———, *Hellroaring: The Life and Times of a Fire Bum.* Saint Cloud, MN: North Star Press, 1994.

John Lyons, *Fire.* New York: Freeman, 1987.

Gerstle Mack, *1906: Surviving San Francisco's Great Earthquake and Fire.* San Francisco: Chronicle Books, 1981.

Norman Maclean, *Young Men and Fire.* Chicago: University of Chicago Press, 1992.

Stephen Porter, *The Great Fire of London.* Gloucestershire, UK: Sutton, 2001.

Stephen J. Pyne, *Introduction to Wildland Fire: Fire Management in the United States.* New York: Wiley-Interscience, 1984.

S. Tahir Qadri, ed., *Fire, Smoke, and Haze: The ASEAN Response Strategy.* Manila, Philippines: Asian Development Bank, 2001.

Harrison E. Salisbury, *Great Black Dragon Fire: A Chinese Inferno.* Boston: Little, Brown, 1989.

Clint Willis, *Fire Fighters: Stories of Survival from the Front Lines of Firefighting.* New York: Thunder's Mouth, 2002.

Periodicals

Carina Dennis, "Burning Issues," *Nature*, January 16, 2003.

Mike Dombeck, "A Tribute to America's Wildland Firefighters," *Fire Management Today*, Winter 2001.

Timothy Egan, "Era of the Big Fire Is Kindled at West's Doors," *New York Times*, June 23, 2003.

Scott Glover, Jack Leonard, and Matt Lait, "12 Killed in Cedar Fire Had No Warning," *Los Angeles Times*, October 30, 2003.

Glenn Hodges, "Russian Smokejumpers," *National Geographic*, August 2002.

Nick Madigan, "Giant Sequoias Threatened as Hot Weather Fuels Wildfires," *New York Times*, July 23, 2002.

Michael Satchell, Jim Moscou, and Reed Karaim, "The Long, Hot Summer," *U.S. News & World Report*, July 8, 2002.

Steve Schmidt, "Interest Is Rekindled in Role of Fire Lookouts," *San Diego Union-Tribune*, September 2, 2002.

Susan J. Tweit, "The Secrets of Fire," *Audubon*, May 2001.

Adam Vincent, "Drought-Driven Forest Fires Singe Retail Sales in West," *Outdoor Retailer*, August 2002.

Internet Sources

Associated Press, "Wildfire Names Always Tell a Story," AZ Central.com, June 19, 2002. www.azcentral.com.

Patrick L. Barry, "Fighting Wildfires Before They Start," Science@ NASA, August 28, 2001. http://science.nasa.gov.

Jamie Brown, "Engulfed in Flames: Bush Fire Training Survives a Reality Check," Weather-wise.com, 2001. www.weather-wise.com.

Annette Trinity-Stevens, "Wildland Firefighters Burn Calories Like Climbers, Soldiers," *Montana State University News*, September 23, 2002. www.montana.edu/news/1032553665.html.

Websites

The Big Burn of 1910, www.missoulian.com. This special-report site links to accounts from survivors of the massive firestorm that burned Montana and Idaho on August 20, 1910.

Firewise, www.firewise.org. The Firewise website offers information about reducing the risk of property damage and loss due to fire.

National Interagency Fire Center (NIFC), www.nifc.gov. This website contains current information on wildland fires, wildfire prevention, statistics, safety, science, and technology, as well as links to the other federal agencies that work together to prevent and fight wildfires in the United States.

National Oceanic and Atmospheric Administration (NOAA) Fire Events, www.osei.noaa.gov. This site makes available NOAA's powerful satellite photographs of major wildland fire events in countries around the world, from Laos to Venezuela.

The Official Website of Yellowstone National Park, www.nps.gov. This National Parks Service website describes the history of fires in Yellowstone and how plants and animals have adapted to survive them.

So You Want to Be a Firefighter?, www.wildfirenews.com. Those interested in wildland firefighting will find plenty to peruse on this site, including information on where and how to apply for careers in wildfire control.

INDEX